Natural Environment Research Council

Institute of Terrestrial Ecology

Annual Report 1978

ISBN 0 904282 31 7

Institute of Terrestrial Ecology
68 Hills Road
Cambridge
CB2 1LA
0223 (Cambridge) 69745

CAPTIONS FOR COVER'S PLATES

The cover shows from the top: White admiral adult butterfly,
photo: E. Pollard; *Mycena galopus* fungus fruiting on oak leaf
litter with signs of white rot caused by this fungus, photo:
J. K. Adamson; the ant, *Tapinoma erraticum*, photo: A. Abbot;
head of female sparrow hawk, photo: I. Newton; the moss,
Sphagnum fallax, photo: R. Daniels; the pseudoscorpion,
Dendrochernes cyrneus (L. Koch), the largest and rarest
British species, on decaying oakwood, photo: J.L. Mason,
Nature Conservancy Council.

Centre: grouse moor, photo: G. Wilson.
Right: an artificial wear machine being demonstrated by
   Alan Frost. The studded rotors used in the field
   experiment have been replaced by rubber covered ones
   in this photograph, photo: A.J.P. Gore.

ACKNOWLEDGEMENT

The Institute wishes to thank Miss Sally Knight for drawing the
figures in this Report. The work was carried out as part of her
year's sandwich course at our Monks Wood Experimental Station,
Huntingdon. Sally is a cartography student at the Luton College
of Higher Education, Bedfordshire.

The Institute of Terrestrial Ecology (*ITE*) was established in 1973,
from the former Nature Conservancy's research stations and
staff, joined later by the Institute of Tree Biology and the Culture
Centre of Algae and Protozoa. ITE contributes to and draws upon
the collective knowledge of the fourteen sister institutes which
make up the *Natural Environment Research Council*, spanning all
the environmental sciences.

The Institute studies the factors determining the structure,
composition and processes of land and freshwater systems, and
of individual plant and animal species. It is developing a sounder
scientific basis for predicting and modelling environmental trends
arising from natural or man-made change. The results of this
research are available to those responsible for the protection,
management and wise use of our natural resources.

Nearly half of ITE's work is research commissioned by customers,
such as the Nature Conservancy Council who require information
for wildlife conservation, the Forestry Commission and the
Department of the Environment. The remainder is fundamental
research supported by NERC.

ITE's expertise is widely used by international organisations in
overseas projects and programmes of research.

Contents

4

The Research Strategy for the Institute of Terrestrial Ecology

Background

The Institute of Terrestrial Ecology (ITE) was created as a consequence of two major decisions. The first, by Parliament, was that statutory responsibility for wildlife conservation in Great Britain should lie with the Nature Conservancy Council (NCC) established in November 1973 as an independent body whose members are appointed by the Secretary of State for the Environment. The second decision was that research on terrestrial and fresh water ecology should remain the responsibility of the Natural Environment Research Council (NERC). Accordingly, the research scientists of the former Nature Conservancy, together with the Council's Institute of Tree Biology, were fashioned into the Institute of Terrestrial Ecology. These new arrangements were compatible with, and indeed might have been made inevitable by, Lord Rothschild's report on Government-supported research and development, the NCC being a customer for ecological knowledge to guide its policies, and ITE within NERC having contractor responsibilities. Subsequently, the Culture Centre of Algae and Protozoa which had formerly been an independent institute, and the Bryology Project Group transferred from the British Antarctic Survey were brought into ITE.

The early years of ITE, from 1973 to the present time, have largely been concerned with the integration of the research activities taking place in each of the research stations. A series of Divisions and Subdivisions has been created within the Institute, each with a leader to guide and develop particular branches of ecological science over the Institute as a whole. These Divisions and Subdivisions are permanent, and provide the necessary intellectual discipline and stimulus for the individual members within them. Since 1973, the whole programme of research has been reviewed, particularly with regard to the objectives of projects and the methods used, and new projects have started to support the research strategy which is described here.

ITE has an integrated programme of ecological research which spans from the fundamental to the applied and covers a wide range of disciplines. A simple, but effective, method of project control and management has been installed to ensure that the research carried out by ITE provides "value for money". Particular attention has been paid to the underlying methodology of the research, developing objective approaches and using quantitative methods wherever appropriate.

External factors

Any development of research nowadays necessarily takes place against a background of financial constraint.

This constraint emphasises, perhaps more than ever before, the importance of good research management to ensure that scarce resources are wisely and economically used, that the research concentrates on correctly assessed priorities, and that full advantage is taken of the best features of the Institute. In particular, it is important to ensure that there is the highest possible collaboration between the sister Institutes of NERC, other research organisations (e.g. ARC, SRC, AERE, FC), and the departments of universities, so that the programmes of research complement each other wherever possible.

Commissioned research is of the greatest importance in ensuring that the scientists of Research Council Institutes establish the scientific basis for environmental policies at regional and national levels. Government departments and other agencies, as customers, define the objectives of research which they would like to have done, and it is for the Research Council, as contractor, to determine the ways in which these objectives can best be met. Ideally, of course, the Research Council Institutes should be able to respond quickly to requests of departments and agencies. Basic or fundamental science, therefore, has a particularly important role to play in providing an adequate theoretical and practical base which will increase the capability and versatility of the Institute to respond to the applied work needed by departments.

Wildlife conservation remains an important aspect of the work of ITE, but ITE has also had to develop a research strategy aimed at providing the scientific base for the application of ecology to the management of the nation's environmental resources. In the development of this strategy, it is, of course, necessary to work closely with colleagues in wildlife conservation, agriculture and forestry who are themselves well-advanced with similar strategies. In this respect, therefore, ITE's main concern has been the consequences of increases in demand for food, fuel and fibre, together with the associated questions of housing, transport, and recreation. The interaction between these economic needs and wildlife, in its widest sense, has necessitated a marked reorientation of ITE's research for the next 5–10 years.

It is, perhaps, worth emphasising the opportunities which are currently being created for research of the kind that ITE now undertakes both at home and abroad. At home, increased emphasis on land use strategies has been focused on the ecology of the land under consideration to ensure that proposed uses are related to the capabilities of the component ecosystems and to the continued maintenance of our soil resources. Furthermore, increasing concern with the environmental impact of industrial and commercial processes has forced attention upon the ecological systems which will need to absorb that impact. Proposals for the development of nuclear and conventional power stations, for example, should take into account adequately the changes that may take place in ecological systems, sometimes at considerable distances from the development, and the fate of various substances that enter these systems. Similarly, our control of pests and pathogens necessarily depends upon increased ecological knowledge of the ways in which these species interact with their environment, and with the species with which they compete. The rabbit, the grey squirrel, and the urban fox all provide examples of animals which are in this particular category, while various aquatic weeds, and Dutch elm disease, provide examples of plants and pathogens.

Economic and social influences

There are several economic, social and political factors which have a major influence upon the priorities set by society on the goals of scientific research. These influences have helped determine the research strategy of ITE, not least by delimiting the types of research for which money is readily available.

Globally, the principal impact on the rural environment during the remainder of this century seems likely to be the increasing demands for fuel, food and wood fibre to meet the needs of the growing world population, and of an improved standard of living for the underdeveloped and developing nations of the world. As fossil fuels become increasingly depleted, alternative energy strategies will need to be derived to make better use of sources of renewable energy and to prevent the destruction of ecological systems upon which sources of renewable energy depend. At the same time, shortages and the increased cost of food may be expected to make radical changes in patterns, forms and impacts of agriculture in relation to the ecology of the rural environment. The marked changes in the cost and availability of the energy inputs to agriculture will themselves generate complex effects on the ecology of land taken into agricultural production. Similarly, a world shortage of timber and wood fibre may be expected to stimulate an increased interest in forestry and a review of the economics of forestry, especially as a renewable source of energy and as a source of raw materials for which recently-developed substitutes (e.g. plastics, synthetic fibres) may become either unavailable or too expensive.

The concern for the rural environment may, therefore, be expected to shift from the current expression of urban man's view of the countryside as a place for recreation and visual amenity to an increased understanding of the countryside as an important source of our basic resources, in which the productivity of our

ecological systems, and the extent to which that productivity can be modified, are the key factors. A similar change can be expected to take place in nature or wildlife conservation, in that the emphasis will shift from conservation as an amenity to conservation as a practical necessity, and ITE expects to strengthen the scientific basis of this change of emphasis in collaboration with NCC. Wildlife organisms, as the component entities of ecological systems, are too important to be treated as a 'residual benefit' of options derived from combinations of other strategies, and research on wildlife organisms is essential to maintain a genetic resource, as alternative species for use by future generations, and to test the resilience, adaptability, sensitivity, and productivity of different species.

Although many forms of pollution are currently declining as a result of improved knowledge and standards, the effects of pollution continue to be one of the principal impacts of man on the ecology of the rural environment, often in unforeseen ways. The ecological pathways and synergistic effects of different pollutants, therefore, demand increasing and more complex research.

Overseas, there are many opportunities in which ITE can help to ensure that developments are consistent with the ecological capabilities of the countries concerned, and in the integration of the complex research that is sometimes necessary for the understanding of ecological problems. While, in the past, ITE has been primarily concerned with research in this country, the many opportunities overseas suggest that a larger proportion of our effort should, in the future, be devoted to problems of other countries, and ITE is already undertaking work for overseas countries, in addition to contributing to the programmes of the Ministry of Overseas Development.

ITE is one of a family of Research Institutes within NERC. It is, therefore, important to ensure that as many of the environmental and land use policies as possible are looked at in consultation with colleagues in other Institutes, and in the universities. While some limited problems can be confined to terrestrial ecology, worthwhile problems for attention by an Institute like ITE increasingly involve such disciplines as hydrology, limnology, atmospheric chemistry, geochemistry, soil science, cartography, mathematics and computer science. We can, therefore, only play our full role by extensive and wholehearted collaboration and integration with our colleagues.

RESEARCH OPPORTUNITIES

Ecological characterization

Ecological survey and mapping of the land use, habitats, organisms and resources of the terrestrial environment are an urgent requirement as a basis for environmental policies, as a base-line for the monitoring of change in ecosystems, and as a means of integrating existing knowledge. Nevertheless, very considerable areas of Britain and other parts of the world have received extraordinarily little ecological attention. Research and detailed investigation have usually been confined to those parts of the terrestrial and fresh water environments which are particularly "interesting" because of the presence of rarities, difficulty of access, or the exhibition of distinctive processes. The remainder of the country, in which equally important changes may be taking place, has therefore been accorded relatively little attention.

The aims of ecological survey may be defined as follows :—

(i)   Provision of sufficient information to provide an effective basis for the characterization of the ecological communities, processes, and dynamics.

(ii) The integration of widely disparate information of environmental characteristics, and the response of plants and animals and the communities they form to these characteristics.

(iii) Interpretation of the inter-relationships between plants and animals and between these organisms and the physical environment, including modifications brought about by man.

Much of the information already available about the terrestrial environment is incorporated in maps at various scales. These maps include information on topography, physiography, geology, soils and vegetation. A first step, therefore, has been to make these data more readily available and more readily utilizable for future work, through modern methods of handling and storing information. Because of the large number of variables and attributes already recorded on maps, examination of the essential correlations between these variables and attributes is necessary in order to determine the nature of the description that they provide, and to establish a basis for the selection of key variables or attributes. A preliminary step has, therefore, been the careful synthesis, aggregation, and analysis of existing information, within the Terrestrial Environment Information System which has been established at ITE's Environmental Data Centre at Bangor.

However, just as ecological survey can be improved by more effective design, the results of ecological survey can now be used in ways that have only recently become possible through the development of more advanced mathematical and computing techniques. We can, therefore, make more intensive use of the information already available, and ensure the integration of any new data with those which already exist. Redundancy of information, correlations between variables and attributes, and the characteristics of ecological variation can all be explored systematically

in the development of models of spatial variation. These models can then be used as a basis for the study of dynamic variation, for example as base-lines for environmental and ecological monitoring, as the framework for Markov models, and in defining the objective function and constraints for linear and non-linear programming. We are already seeing far more analysis of the data collected in surveys, including the determination of the relationships between site factors and the occurrence of different species and assemblages.

There is also a strong link between the ecological survey and characterization developed by ITE and the skills developed by new forms of experimental cartography. Cartography is a necessary part of the presentation of the results of survey and its analysis, and, increasingly, there are considerable benefits to be gained from the methods of digitizing, data storage, display, and plotting now being developed, and particularly from the interactive techniques which enable errors to be edited from maps by successive and progressive stages. The same techniques, however, help to provide information to be incorporated in ecological surveys, and greatly facilitate the establishment of an adequate sampling frame for spatially-related data and the selection of samples representative of defined populations. As these same techniques are increasingly coupled to imagery available from remote sensing and aerial photography, field survey itself has been made at once more rigorous and more closely related to the solution of problems which have themselves been rigorously defined.

Taxonomy and distribution of organisms
Much of ITE's research depends upon the accurate identification of plants and animals, and, therefore, on taxonomic skills at the generic and specific levels. Some problems, including those concerned with forest trees, require taxonomic expertise below the species level, for example in the examination of provenances of introduced trees. Records of the distribution of plants and animals, as in the Biological Records Centre's mapping schemes, also require a sound taxonomic base. These taxonomic abilities are being maintained and developed within ITE as a whole.

Among plants, particular attention has been given to the more difficult groups, including bryophytes, fungi, the rarer flowering plants, and trees. Taxonomy of many species of plants is considerably complicated by the existence of cultivars and clones which may have quite specific adaptability to unusual conditions, resistance or susceptibility to pathogens and pests, or importance as a genetic resource. Similarly, geographically separated populations of some bryophytes have been evolved as a result of vegetative propagation and fungi are able to evolve clones by what is known as heterocaryosis.

Among animals, particular attention has been given to the invertebrates, and perhaps especially to the microfauna, including flagellates and protozoa. The taxonomy and identification of vertebrates below the species level also continue to be of special importance, as has been demonstrated recently by ITE research on such animals as deer and squirrels. Thus, within ITE, it will be essential to retain the skills for the identification of a broad range of organisms, and the capability of specialist taxonomic research.

The study of taxonomic variation of organisms is itself an important field of research, as it is only in the study of such variation that the plasticity of organisms to environmental changes can be determined. Thus, as a range of environmental conditions is created by management and by experimentation, the adaptation of organisms to the changing environmental conditions provides an important area of practical research which can only be successfully pursued through the equally critical study of the genetics, physiology, morphology and behaviour of the organisms themselves. Indeed, much of ITE's research provides valuable opportunities for studies of sub-specific variation in both plants and animals. These studies already include the use of morphometric and chemical techniques, as well as the more conventional methods of taxonomic description.

Physiological ecology
The study of those physiological, genetic and behavioural processes which determine the way individuals and populations live in the wild is basic to much of ITE's work on population ecology and pollution. This observation is particularly true for animals, and a fundamental theme of the work of the Subdivision of Animal Function is that animals undergo marked seasonal and diurnal cycles of body function and nutrition which affect their performance and reaction to environmental hazards such as pollutants. These physiological cycles, affecting such processes as fat storage, moult, reproduction, protein synthesis, etc., are ecologically adaptive and they are controlled by hormonal cycles which appear to be entrained by the photoperiodic clock. Studies of the underlying mechanisms controlling these physiological rhythms are possible now that sensitive and specific radio-immunoassay techniques are available for identifying and monitoring circulating hormones in the blood of birds and mammals. ITE now has excellent facilities for this type of work.

Diurnal and seasonal cycles affect the manner in which animals are exposed to environmental pollutants and the way in which they cope with these chemicals. For example, fat-soluble chemicals, such as chlorinated hydrocarbons, may be stored or released into the blood stream as fat stores are accumulated or used in response to daily and seasonal needs. There are similar affinities between some heavy metals and proteins which may also be affected by cycles in an animal's physiology. Studies on these natural rhythms are fundamental to much of our future work on toxic chemicals.

Work on animal behaviour is paralleled in the Subdivision of Plant Biology by analyses of pheno- and geno-typic responses. How do plants and microbes respond to changing environments, and how have these differing environments historically selected and sorted ranges of within-species variants, ecotypes? These interests are central to our ecophysiological research, where attempts are being made to identify (a) strains of symbiotic microbes able to enhance the growth of their hosts in a variety of situations, and (b) ecotypes among bryophytes, grasses and woody perennials, including trees, with differing patterns of growth. This work is focused on events in temperate and polar regions, but is already being extended to the tropics, e.g. to help conserve species of tropical hardwoods.

By providing an understanding of plant and animal responses to different environments, ecophysiological research enables predictions to be made that may influence management procedures. It is likely that studies of reproductive and vegetative strategies will contribute to the conservation of rare plants, (e.g. *Gentiana pneumonanthe*), to the identification, and hence selection, of improved strains for stabilizing mudflats and disturbed and exposed sites in the uplands, for withstanding trampling in areas subject to recreation pressures, and to the exploitation of "superior" crosses and clones of timber trees which may, in the future, become a major source of fuel.

Population processes
The study of the population dynamics of individual species has long been a strength of the Institute. Besides leading to a better understanding of the reasons why some species are abundant, others are rare, and why some populations are more stable than others, work in this field has helped to elucidate the role of spacing behaviour, aggression and nutrition in population processes. Furthermore, the study of population ecology is fundamental to solving many applied problems.

In the past, much of ITE's research on animal populations has been on organisms of particular importance for wildlife conservation. Some work on rare or endangered species will undoubtedly be required in the future, but the expertise within the Institute is now being used to tackle a far wider range of problems than those associated with conservation. All of ITE's work on population ecology is aimed at population management, i.e. the manipulation of the environment in order to maintain populations at required levels. In the case of pest species (such as the rabbit, grey squirrel and *Scolytus intricatus*), new work is aimed at reducing damage by ecological means, whilst for exploited species (e.g. red grouse, Roman snail) and rarities (e.g. the large blue butterfly and otter) the studies aim at maintaining populations at a viable size. Increasing emphasis is being given to the population dynamics of potentially useful species of herbivores (e.g. sheep and deer) and the interaction with plants and plant processes.

From its inception, ITE has had a strong interest in the description of plant assemblages which has been augmented by dynamic approaches to community ecology, including sequential surveys (monitoring) and series of experiments. Our plant population studies are concerned with the ability of plants to survive when competing with others of the same or different species, or when responding to the damage done by herbivores, by pathogens, by trampling, etc. These studies bring together colleagues concerned with different aspects of grazing, not only of herbage but also of phytoplankton (by zooplankton), with insect pests and vermin restricting plant regeneration and/or the growth of forests and managed woodlands, and with others evaluating the effects of past management on plant successions. Increasingly, there is an awareness of the potential value of the history of the use and management of land as a tool to help explain present-day differences between plant assemblages. In many instances, it enables timescales to be attached to a series of phased events which, coincidentally, may profoundly affect soils.

Ecosystem studies
The investigation of basic ecological processes is fundamental to our understanding of the ways in which ecosystems operate and are modified by environmental changes, including those initiated by man. Research needs to be focused on the functioning of ecological systems as a whole through the development of long-term, multidisciplinary studies carried out in geographically widely-separated localities to ensure that the full range of variability in space and time is encompassed by the investigations. The study of primary and secondary production, which formed the basis for much of the work of the International Biological Programme, now needs to be extended to a more fundamental study of the ways in which these processes contribute to change, resilience and stability of ecosystems. Improved knowledge of the physiology and the genetics of wildlife organisms is a principal requirement for the increased understanding of ecological systems necessary for the rational management of the terrestrial environment and the prevention of damage or loss of function. An increasing emphasis on the need to produce food, fuel and fibre will place a heavy burden upon natural and semi-natural systems, and this burden needs to be anticipated if we are to avoid attempts to make demands upon ecological systems which may well end with their destruction. In addition, we need better understanding of the consequences of the destruction of natural and semi-natural systems by the imposition of crop systems of different kinds.

Plant communities should not be studied in isolation from their habitats. In the past, physiologists have tended to concentrate on the interception and exploitation of solar energy by foliage, but, with improving techniques, the intricacies of root growth are also being studied. How is root development related to the availability of water and nutrients, the latter in natural and semi-natural situations being cycled as a result of microbial activity? Much still needs to be learnt about the cycling of phosphorus, nitrogen and sulphur and of the effects of pollutants and different management practices, including clear-felling and re-afforestation, upon them. This information is required to enable our soil resources to be sustained, a task requiring a vigorous group of soil scientists studying not only the effects of soil on plant growth, but also those of plants on soil properties, physico-chemical and biological.

The monitoring and surveillance of dynamic change in ecosystems have increasingly been identified as an essential component of ecological research which has been neglected because of the difficulties of supporting research of this sort for long periods. The UNESCO Man and the Biosphere (MAB) programme and the Scientific Committee on Problems of the Environment (SCOPE) have both stressed the need for emphasis on the measurement of dynamic change in ecological processes. While many of the research priorities discussed above may be regarded as setting base-lines for monitoring and surveillance, particular attention is being given to those aspects of ITE research which will more directly develop an understanding of dynamic change itself.

For plants, the determination and mapping of changing distributions and populations have already been given some priority in ITE. The distribution maps of the Biological Records Centre, for example, have provided a stimulus for active research on many species and a basis for the Red Data Books. However, intensification of research on changes in plant distribution and populations (together with their associated animals) can be expected to form an important component of future ITE research, but with careful selection of species that seem likely to relate to key ecological principles identified by the research on ecological processes. More important, however, is the development of research on dynamic change in plant ecosystems, the identification of seral stages leading to climax or sub-climax vegetation, and the effects of human modification of ecosystems. In addition to the techniques being exploited by historical ecologists, others now exist for the study of such changes, and the main difficulty is likely to be the availability of research areas which can be maintained for sufficiently long periods of time.

The wide variations in climatic conditions in Britain, and the unresolved conflicts between experts on the causes and predictions of long range climatic trends have focused attention on the need for more precise information on the response of ecological systems to climatic changes. Almost no attention has been given to the establishment of experimental plots from which ecological responses to climatic change can be determined, and this omission will be remedied by ITE for the future by ensuring that parallel meteorological and phenological records are available. ITE is one of the very few Institutes in Britain with the expertise and resources to establish such experiments and maintain them for sufficiently long periods of time. Some fairly rapid progress with this research may be achieved by the identification and analysis of critical phenological variables, but other methods and observations will be needed to identify the climatic factors which have the greatest impact on ecological systems and to determine the ways in which these factors interact and are integrated within the system.

Land use

All the aspects of ecological research which have been mentioned so far build heavily on the traditional development of the ecology of systems which are not intensively cropped. A distinctive feature of future research in ITE is an increasing concern with the effects of human management within the rural environment. Expressed in this way, our concern is not with comparisons of relatively small changes in agriculture and forestry, but with the longer-term effects of intensive agriculture and forestry—incorporating trends towards more extensive systems—by comparison with those systems of management which have been traditionally employed. Such research is necessarily based on long-term studies and experiments which test and compare combinations of land use policies and techniques.

ITE research in this field is, therefore, relevant to many important aspects of agriculture, forestry, wildlife conservation, amenity, and water resources. While some emphasis and interest is inevitably focused on the effects of past and current changes on ecosystems, the main aim is to develop a predictive capability to guide future policies. For example, the likely effects of changes which will be induced by the increasing cost of energy from fossil fuels need to be anticipated if agricultural and forest policies are themselves to be modified in ways which will be beneficial both to the environment and man. Similarly, the Nature Conservancy Council has already expressed its concern about the effects of the changes which have already taken place in agriculture, and the likely effects of further changes now need to be predicted if policies for wildlife conservation are to be effective.

In addition, there are some particular problems to be addressed within the general ecology of Britain. One of the most urgent of these problems is that of derelict land—a land use category which needs more careful definition as a first part of the research. Apart from this need for improved definition, the assessment of basic fertility and its restoration is an essential component of the research which needs to be taken to a reasonable stage before any predictions can be made of the likelihood of success in restoring such areas to agriculture, or to some form of forestry, conventional or non-conventional. Erosion, frequently induced by such treatments of existing problems as the use of chemicals for the control of bracken, or by recreation on sand dunes, is another problem requiring early solution in many ecosystems. The dry summers of recent years have also emphasised the hazards of fire when ecological systems are allowed to accumulate a considerable biomass, the difficulties of reclamation of fire-damaged systems, and the management of ecosystems to minimize the incidence and severity of fires.

Because there are inevitably many institutions, organisations, and agencies interested in the effects of their management and policies, ITE's distinctive orientation perhaps needs to be stressed. It is not the concern of ITE to concentrate its research on management effects on individual sites or individual ecosystems for their own sake. Rather, the aim is to integrate research on several sites to provide information about regions of Britain or for national and international syntheses. ITE's role is, therefore, to provide a framework wherever feasible for the separate strands of research of other agencies, filling in, where necessary, with its own experiments and surveys. In this co-ordination, the development of systems analysis and modelling, discussed in a later section, will be an essential carrier for the linking of the more direct surveys and experiments.

Fresh water habitats

The primary development of ITE's fresh water expertise is linked to the interface between fresh water and terrestrial ecosystems. In particular, changes in land use and in the technology of agriculture and forestry may be expected to have major impacts on the hydrology, physical and chemical properties, and biology of fresh water systems, and these impacts and changes should be anticipated and predicted, and taken into account in the decisions about future land use. ITE's special concern is for the pathways of elements and compounds, e.g. heavy metals, phosphates, nitrates, sulphates, etc., and the response of plant and animal organisms to variations in these substances.

The fate of pollutants and pesticides applied directly to fresh water systems, or applied to terrestrial systems but ultimately reaching fresh water systems and organisms, can be expected to be a continuing priority, together with fundamental research on the ways in which organisms are tolerant or intolerant of these substances. Some organisms act as convenient early warning indicators of accumulations which may become hazards to human health, but increased knowledge of the biological pathways is vital to our understanding of the ways in which our ecology and environment may be modified in unforeseen ways. ITE's development of a continuous flow system is the first stage of an approach to this important problem which will be further developed as we gain expertise and experience.

Aquatic herbicides, already under investigation, may also be expected to be a focus for continuing research, together with research on the fate of nitrate and phosphate fertilizers.

Land/water interfaces
Ecological studies of land and water bodies need to take into account the special conditions that arise at land water interfaces. Such situations are not merely intermediate, they are distinctive. For example, the fate of pollutants entering water bodies (fresh water or marine) is influenced by interactions at the very large aggregate surface area presented by suspended particulate matter of complex organic and inorganic origins. Studies at both the macro and micro levels of land/water interfaces are essential for a full understanding of pathways of nutrients or pollutants, including radionuclides.

Systems analysis and modelling
It scarcely needs to be said that ITE's research is firmly based on the valid design and efficient analysis of experiments and surveys. Statistical theory and practice are both in the process of rapid development, and ITE maintains sufficient expertise to follow this development, and to contribute to the development, even though statistical research itself is not a primary objective.

There is, however, a complementary area of development in the systems analysis and modelling of ecosystems, namely the synthesis of ecological theory into deterministic and stochastic mathematical models. Many, if not most, of these models will be used in the simulation of ecological processes and ecological systems. Such simulations may be used to predict the likely effects of management policies on the stability and resilience of ecosystems, a goal which the majority of resource managers have yet to find acceptable—but they may also be used to test the conformity of existing theories, to test the sensitivity of our perceptions of ecological relationships to small changes in the parameters of those relationships, and to test formal hypotheses against the model systems. If this systems analysis and modelling can be carried out in anticipation of the emergence of critical problems, the influence of ecologists on the management of natural resources can be greatly increased.

As our understanding of ecological systems, and the mathematical modelling of those systems, improve, it may increasingly become possible to switch the emphasis from simulation models to decision models, i.e. to make models which aid decisions by indicating ranges of options which are optimum in some defined way. Initially, as at present, much of the emphasis of these decision models is on static, one-off problems. ITE, is for example, currently working actively with several agencies on a linear programming model of

Cumbria and other areas. With greater experience, such models may well become more dynamic, ensuring that resource management decisions taken at one time do not close options that need to be maintained for future decisions. The explicit development of systems analysis and modelling as a research objective of ITE represents an orientation which gives the Institute a distinctive role in the field of terrestrial ecology.

Although statistical and mathematical research is not a primary objective of ITE, we nevertheless have an important responsibility for the practical development of systems analysis and modelling as applied to ecological problems, the description of applications to such problems, and the formation of the general philosophy. This responsibility is partly national, within NERC and within the professional fields of ecology, and partly international. Indeed, ITE already contributes to the programmes of MAB and SCOPE through its expertise in systems analysis, and has substantial contracts from these organisations.

Conclusion
ITE, as one of the component Institutes of the Natural Environment Research Council, represents a major Research Institute for terrestrial and fresh water ecology. The expertise and resources of the Institute are considerable and are already being focused on problems of national and regional importance, much of the research being commissioned by international and national agencies. The importance of ecological studies can only increase as our awareness of future problems itself increases, and as we seek to develop policies for natural resources in the developed and developing world. ITE's research is intended to gain relevant ecological knowledge before our impact on the ecological systems of the world has undesirable and unexpected consequences.

Many international and intergovernmental agencies have programmes and projects for terrestrial and fresh water ecological research, including the United Nations Educational, Scientific and Cultural Organisation's Man and the Biosphere programme, the working parties of the Scientific Committee on Problems of the Environment, the United Nations Environment Programme, the International Institute for Applied Systems Analysis, and the International Union for the Conservation of Nature and Natural Resources. ITE already has links with many of these programmes, and will seek increasingly to create further opportunities to develop the expertise of its staff and to provide advice and research overseas, as well as in Britain. Science is essentially international, and the science of ecology, in particular, transcends national and cultural boundaries as man seeks to find ways of living in harmony with his environment.

J.N.R. JEFFERS
Director, ITE.

Longer Research Reports

Introduction
This section of the report contains descriptions of research which has been completed or has reached a stage justifying rather longer reports than those contained in section III.

The first of these reports is a brief summary of a desk study carried out by ITE on the distribution and movement of radionuclides in terrestrial ecosystems. This represents a new field of research for ITE, but one which is expected to develop rapidly during the next few years. Field studies of the processes and organisms involved in the transfer of radionuclides along environmental pathways have now been planned.

Studies of the effects of environment on forest growth begin to assume increased importance as the likely future shortage of timber and wood products comes to be realised. ITE's research on the influence of environmental factors is intended to complement research on the genetics of forest trees. The results so far achieved, and the implication of those results are summarised in this report.

ITE has been involved in work on amenity grasslands for some years, and, in this report, there is a summary and discussion of the results of a series of experiments on the effects of wear and nitrogen on sports turf produced from various mixtures of grass species.

A rather different kind of report summarises the results from some long-term fundamental research, first begun in one of the projects of the International Biological Programme. As a result of the research, we are now able to assess the activities of one of the key organisms in woodland soil decomposition.

The next two reports deal, respectively, with survey of the invertebrate fauna of dune and machair sites in Scotland, and with research on the interaction between oystercatchers and mussels. The two reports illustrate the wide diversity of ITE's research in animal ecology.

Finally, the section ends with a report on the progress made in research on cryopreservation of algae and protozoa. Sufficient progress has been made to suggest that it may be possible to store a range of organisms without reculturing in the relatively near future.

RADIONUCLIDES IN TERRESTRIAL ECOSYSTEMS

Introduction
The majority of radioactivity discharged by the nuclear industry in the UK is released to aquatic ecosystems. Some of this discharge will be transferred eventually to land, adding to the radioactivity already present from natural sources and from fall-out from nuclear explosions and atmospheric discharges by the nuclear industry.

In 1976, the Royal Commission on Environmental Pollution concluded that insufficient research had been undertaken on radioactivity in the terrestrial environment and that independent research in this field by the Research Councils should be encouraged. These points were noted in a Government White Paper in 1977 and taken into account subsequently by the Department of the Environment in a review of research requirements. Subsequent discussion between ITE and DoE led to a desk study which summarized the available information on the distribution and movement of radionuclides in terrestrial ecosystems, with particular reference to north-west England, and which identified the gaps in our knowledge.

Transfer of radionuclides
Transfer pathways identified during the desk study are indicated in figure 1. Radionuclides may enter terrestrial ecosystems by several pathways: in precipitation, as particles and dissolved gases or vapours; in dry deposition, as particles or aerosols emitted by the nuclear industry or resuspended from soil, vegetation and water surfaces; in mineral matter and seaweed deposited on the seashore and around fresh water courses and lakes; in birds or invertebrates which feed on marine organisms, but which also periodically move inland; in fish or shellfish consumed by man and other animals; in material removed from the shore by man, e.g. seaweed as fertilizer and sand; in fertilizers especially phosphate; and by burial of radioactive waste.

Radionuclides deposited on plants and on the soil surface occur in various chemical forms in particles, as vapours, or in solution. Each nuclide may be deposited in more than one chemical or physical form.

Retention by a plant depends on its surface characteristics and the growth form. Wind removes all particles $> 44\,\mu m$ deposited on the plant. From 83–90 per cent of deposited water-soluble radionuclide may be lost from plants in a month, with little difference in loss between plant species or radionuclides. Retention half-life for radioactive aerosols and fine particles in agricultural and non-agricultural vegetation varies from 13–87 days.

Relevant literature indicates that the retention of radionuclides by soil varies with physical and chemical characteristics of the element involved, the physical and chemical form of the nuclide, the concentrations of the radionuclide and of other ions which compete for retention sites, soil pH, soil organic matter content, soil mineral type, concentrations of chelating agents or other compounds which form complexes with the radio-

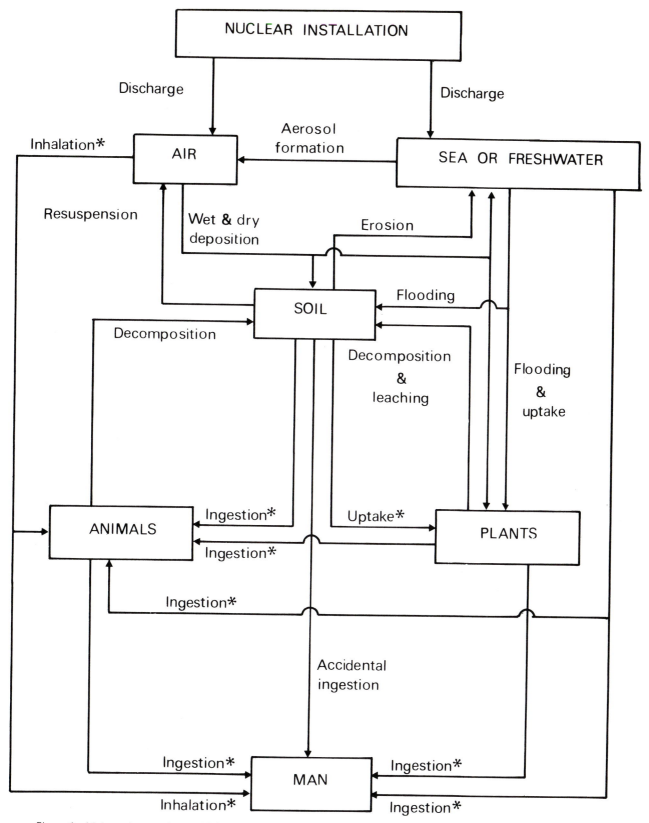

Figure 1   Main pathways along which radionuclides are transferred into, throughout and within terrestrial ecosystems.

nuclide, and the time allowed for equilibration of the radionuclide between solid and liquid phases. Nuclides are retained by soils in water films, in hydrated oxide or humus on particle surfaces, and in the crystal lattice of soil minerals. They occur in solution, as cations, anions or complexes with mineral or organic matter, and as particles. Sr radionuclides are retained by soil mainly in solution, or on minerals or organic matter in an ion-exchangeable form. Cs, Ru, Zr, Nb, Ce and other rare earths and the transuranides are held more strongly than Sr, particularly by soil minerals. Sesquioxides appear to be especially important in retention of I. Both I and Ru can occur in anionic form and are held strongly by acid organic matter. Maximum retention by soils occurs at pHs 7–11 (Sr), 3·5–11 (Cs), 1–5 and > 8 (Zr × Nb), 6–8 and > 11 (Rare earths), 4–9 (Transuranides) and 5·5–8·5 (l). The effect of pH on Ru retention by soil varies with the experimental conditions. Chelating agents affect retention and migration of Sr significantly only in soils that are low in clay. They have little effect on Cs and Ru retention, but promote desorption of the rare earths and transuranides.

The distribution of radionuclides in soil profiles is affected by the retention characteristics indicated above, precipitation, temperature, moisture, soil composition and the degrees of soil fissuring, biological activity, cultivation and fertilizer application.

All parts of plants may absorb radionuclides. Absorption by the aboveground parts appears to have been little studied, although many data are available for the gross retention of 'natural' or simulated fall-out by these parts. These data indicate that 4–75 per cent of radioactivity in plants in the field is derived from direct deposition, the value varying with radionuclide and plant type.

Absorption of radionuclides by plant roots varies with the many factors which affect retention and distribution of the nuclides in soil. It varies by one of several orders of magnitude between plant species, soil types and with radionuclide, according to the following decreasing order $Sr \gg I > Cs \simeq Ru > Ce \simeq Y \simeq Pm \simeq Zr + Nb > Pu$. Some elements, e.g. Cs, are mobile in all directions in the plant whereas others, e.g. Sr, are mobile largely from root to shoot, and yet others, e.g. Zr, tend to remain near the site of absorption. Cs and Sr, consequently, tend to become evenly distributed throughout a plant, whereas Zr concentrates in the roots.

Absorption by plant root varies with the root's distribution in the soil profile. Shallow-rooting species such as ryegrass absorb decreasing amounts of ^{90}Sr over several years as the nuclide migrates downwards from the soil surface, whereas deep-rooting species such as lucerne absorb less ^{90}Sr than ryegrass when the contamination is near the soil surface. Ploughing or soil mixing by earthworms changes the distribution of radioactivity in the soil profile, and hence affects plant uptake. The absorption by plants of some radionuclides, e.g. Pu, increases with time, possibly because of changes in the availability of the radionuclide during the root growth/decay cycle or solubilization of unavailable radionuclides by micro-organisms.

Uptake and retention of radionuclides by animals vary greatly with radionuclide and animal type. Vertebrates and those invertebrates with a calcareous exoskeleton or shell, for example, have a higher demand for Ca and Sr than other invertebrates. Uptake occurs via respiratory surfaces, the skin and the gastrointestinal tract. Elements such as Ca and Sr, which are readily taken up by plants from soil, are readily absorbed in the animal gut whereas others, e.g. the transuranides, are absorbed negligibly. The distribution of absorbed radioactivity in the animal body is usually very uneven and varies with the radionuclide and animal involved. Elimination of radionuclides by the animal is a complex process involving different rates and different parts of the body.

The rate of release of radionuclides during decomposition of plant remains varies with the radionuclide involved. For example, ^{85}Sr and ^{106}Ru are lost at the same rate as dry matter, whereas ^{134}Cs is lost at more than twice that rate because of leaching by rain. Like stable nuclides, radionuclides are immobilized by micro-organisms in plant remains. Retention half-life for radionuclides in decomposing animal remains appears to vary between a day or two for soft tissues to many thousands of years for bones and shells.

Radionuclides are lost from ecosystems to the air as gases, vapours and in resuspended soil particles, to fresh water and the sea in solution or in eroded soil, and to various ecosystems in crops, timber and animal products. Losses of radionuclides from UK ecosystems appear to be unknown. Only 0·35–4·5 per cent of the annual fall-out input of ^{90}Sr was lost annually in eroded soil and solution from some agricultural soils in USA. Some 46–57 per cent of the incoming ^{90}Sr was lost from other agricultural soils. Eroded soil was about 10 times richer in ^{90}Sr than the uneroded soil. A strong association between soil erosion and radionuclide loss has also been found for ^{137}Cs and ^{239}, ^{240}Pu in the USA.

Future work
The desk study and subsequent assessment identified 12 research areas of interest to ITE. Four of these have now been selected for more detailed study on the basis of feasibility and priority in relation to the interests and requirements of other organizations. Emphasis will be placed on processes and organisms involved in the transfer of radionuclides along environmental pathways, and on the importance of various environmental factors and land management and use in this transfer. The studies will, therefore, be largely complementary to the monitoring undertaken by BNFL, AERE Harwell, NRPB Harwell, MAFF and others. Close liaison will be maintained with the organizations to avoid duplication of effort and to share resources and expertise. Appropriate staff, including a radiochemist, are now being recruited or allocated, and relevant equipment and other facilities are being acquired. It is hoped to begin specific field studies in early 1979.

K.L. Bocock

EFFECTS OF ENVIRONMENT ON FOREST GROWTH

The annual consumption of timber and forest products in the United Kingdom is valued at £2500 million, of which only 8 per cent is home-grown. The majority of the UK's 1·7 million ha of forest was planted during the last 25 years and, as this forest enters its productive phase, the UK will produce 14 per cent of its requirements by the year 2000 (Holmes in press). In addition to augmenting the area afforested, amounts of timber produced may be enhanced by accelerated growth rates. It is important that these differing options should be explored. A world timber shortage is predicted for the end of the century (Holmes in press) and it is likely, because of economic factors, including employment, that exporting countries will wish to sell increasing proportions of processed, as opposed to raw, timber.

Using existing criteria, a further 3 million ha in the uplands could be expected to produce rewarding crops of timber. However, restrictions of various sorts reduce even the most ambitious scheme to some 2 million ha. This addition would double the existing area, but still leave the UK with a short-fall of 70 per cent of its timber requirement. There is, therefore, a need to seek methods of increasing biological efficiency, and to determine what scope there is for such an increase. Agricultural productivity has risen steadily for many years by using 'improved' varieties, by changing cultural techniques, and improving methods of crop protection. From an empirical beginning, yields of wheat, sugar beet and potatoes have been increased by 59, 66 and 40 per cent respectively by an increasingly analytical approach (Bingham 1971). Can we expect similar major increases in forest productivity, recognising that forest trees are, in comparison, relatively "unimproved"?

There are good reasons for supposing that appreciable potential genetic gains exist, but they are likely to be more difficult to realise than in agriculture. Many agricultural crops are annuals and the plants themselves are of a size to enable replicated experiments to be made with comparative ease, with consequent crop improvements. In contrast, the variability of upland sites and the extended cropping cycle (rotation) of 50 years tend to make complex forest situations even more complex.

In considering the factors controlling the growth of Sitka spruce *Picea sitchensis* (Bong.) Carr. at a site in the hills near Moffat, Dumfriesshire (plate 1), in the centre of the large belt of plantation forests extending across southern Scotland and northern England, it was, from the outset, recognised that: (i) forests create their own micro-climates, which change as forest structure changes on ageing and (ii) weather patterns in upland Britain can change rapidly (Plate 21).

I *Stand structure and forest micro-climate*

(a) Development of stand structure
Growth rates of young forest plantations increase rapidly, but a limit is reached as the available resources are increasingly exploited. These changes were observed when studying the branch development of young *P. sitchensis*, a permanent scaffold facilitating the measurement of numbers and lengths of branches in successive years. As a result, 3 phases of stand development were recognised (Cochrane and Ford 1978):

(i) An initial period before 'foliage overlap' (between neighbouring trees) when the small annual height increments progressively increased as trees grew larger and exploited the resources of the site at an increasing rate (figure 2).

(ii) A period between 'foliage overlap' and 'crown interlock' when height increments increased more rapidly, probably attributable to the combined effects of (a) the drying out of very wet soil as growing trees intercept increasing amounts of rain, (b) the death of ground vegetation and the possible release of nutrients as a result of shading, and (c) a reduction in numbers of branches produced by trees during this period, so decreasing competition for resources within trees and allowing height growth to be enhanced. During this period, there is no competition between individual trees; they all have similar relative growth rates so maintaining the size differences that existed at planting.

(iii) A period after 'crown interlock' when annual height increments varied around a stable mean, and when competition between individuals occurred with size differences widening as the growth rates of large trees increased relatively to those of small trees.

By year–14, the surface areas of needles (both sides), branches and tree trunks were 26 times larger than the area of ground covered. Branches of neighbouring trees interlocked extensively so that 25, 44, 28 and 3 per cent of the ground area was covered by the crowns of 1, 2, 3 and 4 trees respectively. The main branches which were angled downwards and inwards were arranged in whorls and carried 70 per cent of the foliage.

(b) Drying of forest soil
For most of the year, precipitation exceeds potential transpiration on the moors of upland Britain and soils are frequently wet or water-logged to an extent that tree growth is restricted unless the ground is ploughed. However, measurements, over a year, of amounts of precipitation reaching soil beneath the canopy by 'throughfall' (dripping) and 'stemflow' were 30 per cent less than those assessing amounts of precipitation falling outside the forest (Ford and Deans 1978). This 30 per cent 'interception loss' of water, evaporated directly from foliage and branches, is twice as large as the average value for upland moorland. Interception losses were less in winter (15 per cent) than in summer (40 per cent), when more energy was available to effect evaporation and when rainfall generally occurred in smaller amounts at any one occasion (figure 3).

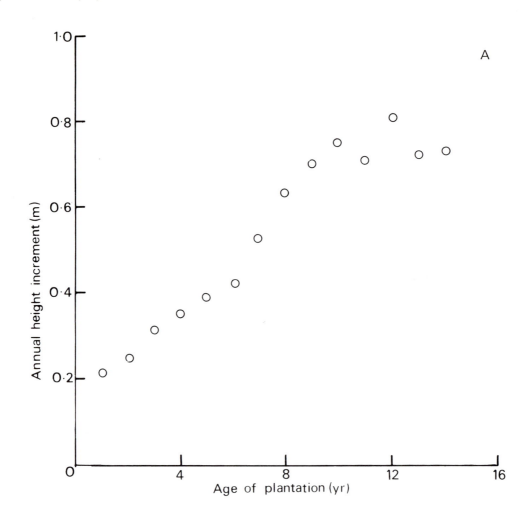

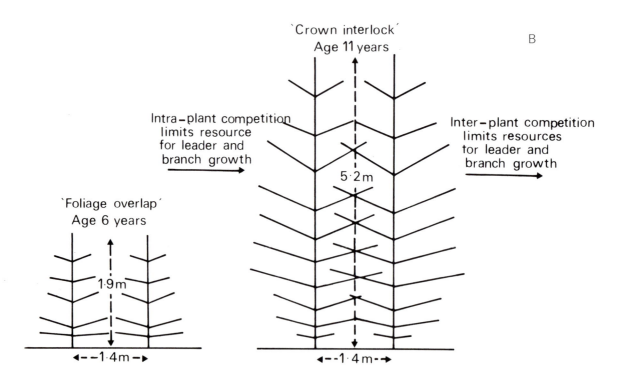

Figure 2 The early growth of young Sitka spruce, (A) successive annual height increments of the leading shoot and (B) diagrams illustrating two critical stages of plantation development (i) foliage overlap and (ii) crown interlock.

Photo: J N Greatorex-Davies

Photo: R C Welch

Photo: E Duffey

Photo: R C Welch

Photo: E Duffey

1	6
2	7
3	
4	
5	

Photo: R C Welch,

Plate 1 Highland Fabricators Yard viewed from hill of Nigg, looking over the Cromarty Firth towards Invergordon.
The construction yard for making oil rigs was built up in the early 1970s. Most of the dune system in the photograph is also scheduled for industrial development, leaving only a small strip of dunes on the left of the road.

Plate 2 Faraid Head (site 56), Sutherland.
One of the largest dune systems on the north coast of Sutherland. This site was surveyed for invertebrates during June and July 1976.

Plate 3 Installation of light trap at Bettyhill (Invernaver) site 57, Sutherland.
The trap was placed in a shallow dune depression in the marram transition zone. The battery was enclosed in a polythene bag and buried in the sand adjacent to the trap.

Plate 4 View looking north of Luskentyre Banks, Harris, Outer Hebrides (site 40).
The only extensive mobile dune system in the Outer Hebrides which was sampled during the 1976 invertebrate survey.

Plate 5 Lycia zonaria male (Belted beauty moth)
This moth is very local in Britain but is widespread in the Outer Hebrides where the survey recorded larvae from 16 out of the 18 sites sampled.

Plate 6 Ormiclate, (site 22) South Uist, Outer Hebrides.
A typical pitfall trap site on the floristically rich machair dunes.

Plate 7 Barvas (site 43), Lewis, Outer Hebrides. A pitfall trap site in short, dense, machair turf grazed by sheep and rabbits.

Photo: H Arnold

Plate 8   *Sparrowhawk at nest in a Dumfriesshire woodland.*   *Photo: I Newton*

Plate 9   *Contrasting growth habits in cuttings from the same clone of* Triplochiton scleroxylon *when taken 29 months after the parent seedling germinated.*
(*a*) Vertical plant, *with radial leaf arrangement, originating from* basal *shoots on stockplant grown at an oblique angle.*
(*b*) Non-vertical plant, *with distichous leaf arrangement, originating from an* apical *shoot.*
Photo: K A Longman

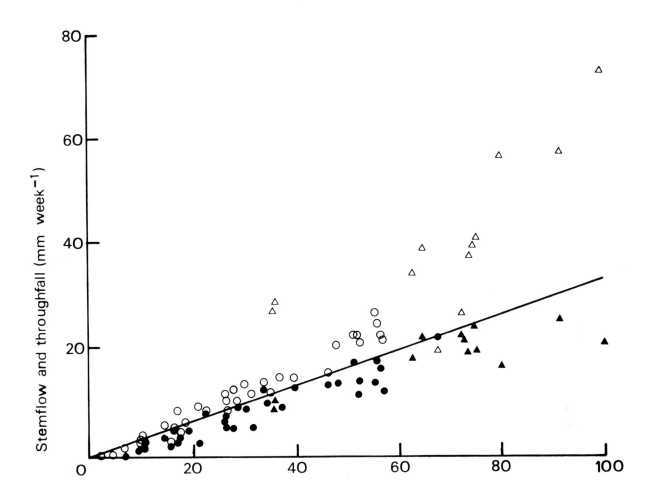

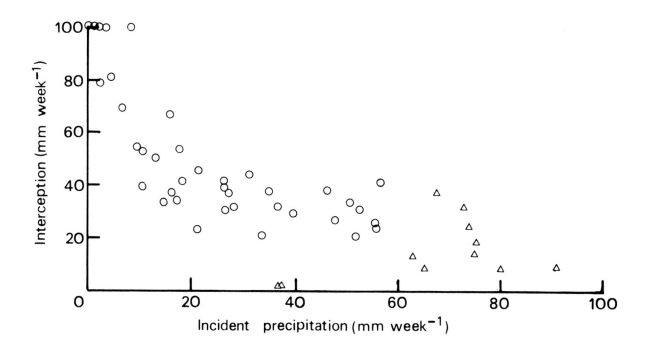

Figure 3 Relations between incident precipitation and (A) stemflow (○, △) and throughfall (●, ▲) and (B) interception losses. A line with a slope of 0.33 is included in (A).

Averaged over the summer growing period, it was estimated that trees received 7·6 litre of water/day. To understand how this amount is used, it is necessary to estimate the amounts lost by transpiration—a technically difficult task. However, they can be estimated indirectly from energy budgets (Milne in press). Having measured net radiation, it was assumed that, during the day, it (a) heated the atmosphere and was transferred as 'sensible heat' and (b) was used as the latent heat of evaporation in the transpirational process. In forests with dense canopies, only a small fraction of net radiation heats soil, and, because of efficient aerodynamic mixing within forest canopies, the energy used to heat needles and other aerial parts of plants is also relatively small.

The transfer of radiation as 'sensible heat' was measured with a 'fluxatron' (Milne in press), an instrument which correlates the upward movement of eddies of air with their temperature to give an integrated value of heat movement over periods, for example, of an hour. It was assumed that the remaining net radiation was used in transpiration, and, knowing the latent heat of evaporation, this transpiration was calculated in mm of water or as an average in litres/tree.

Over a 6-day period with a range of typical summer weather conditions, average transpiration amounted to 3 mm/day, i.e. 6 litres/tree/day. Theoretically, 1·6 litres would be available per tree, but, in reality, this is likely to be a considerable over-estimate because allowances have not been made for surface run-off.

(c) Patterns of soil water and the distribution of fine roots

The distribution of water within the forest was not uniform (Ford and Deans 1978). Forty per cent occurred as stemflow and, of the remainder, throughfall, there were always greater amounts near to the bases of trees and in the gaps between trees within a row than in the areas between rows. These large amounts of water close to mainstem are attributable to (i) large densities of foliage which occur midway between rows of trees and which effect efficient interception and (ii) the erect branching habit of *P. sitchensis* which, when observed (Cochrane and Ford 1978), ensured that water would be conducted towards stems. These attributes of stand structure are likely to change as the particular plantation matures and also when it is thinned. The crowns of individual trees will become more 'open' and, concomitantly, branches which slope outwards will form a larger proportion of the total. When this change occurs, the proportion of stemflow will decrease while throughfall will become more widely scattered (Aussenac 1970).

The concentration of incoming water near stems and in gaps between trees along rows was directly paralleled by the distribution of fine roots within soil. However, in addition to the large amounts of throughfall, the edaphic environment along rows was enhanced by the upturned ribbon of turf left after ploughing, a ribbon increasing the concentrations of available nutrients (Ford and Deans 1977). Thus, it would seem possible that increased moisture availability and nutrient release (Heal in press) together create a more favourable environment for root growth. Because thick roots were found extending considerable distances, it seems that the factors controlling their growth differ from those influencing fine roots which are concentrated near to stem bases.

II *Effects of changing weather on tree growth*

Three aspects of tree growth have been considered—shoot growth, cambial activity and the populations of fine roots.

(a) Shoot growth
In relating daily increments with weather, it was found that shoot elongation was closely correlated with solar radiation; elongation rate increased following sunny days. Additionally, and of possibly greater significance, it was found that the growth of shoots in different parts of the canopy was not synchronous. Not surprisingly, therefore, numbers of generally applicable correlations were severely limited.

Shoot elongation both started and stopped sooner on basal than on upper branches. It started 3–4 days earlier and continued for 50 days: on top branches it continued for 70 days, with leading shoots elongating for 95 days. During the period of shoot extension, daily solar radiation and temperatures gradually increased. It was found that extension of lower shoots, and of the leader, was positively correlated with temperature, an effect detected only because most of their extension occurred when mean temperatures were low. With a newly-devised automatic shoot extension sensor (Milne, Smith and Ford 1977), sensitive to 0·2 mm, 2 effects were detected on sunny days, viz. (i) shoot contraction as water was lost during the day and (ii) accelerated extension after re-hydration in the early hours of the ensuing day (figure 4). The capacity of plants to respond to strong daytime radiation is limited by amounts of available water.

(b) Cambial activity
During the formation of spring wood, cambial activity was positively correlated to solar radiation as found when small segments of stem were sectioned and examined microscopically. Daily esti-

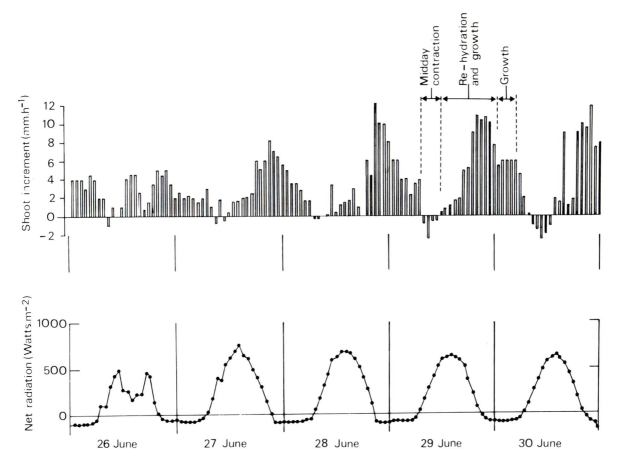

Figure 4   *Relation between hourly shoot extension increments and the diurnally changing amounts of net radiation in a young plantation of Sitka spruce.*

mates of numbers of cells formed in radial files of tracheids (Ford, Robards and Piney 1978) showed that their development was related to radiation the more numerous cells having larger volumes and thicker walls than average (Ford and Robards 1976).

(c) Growth of fine roots
In contrast to the daily records of shoot extension and cambial activity, root growth has been measured at intervals of 5 days using a coring technique, an adequate interval remembering that fluctuations in soil environment are less rapid than those above ground. Populations of fine roots in soil are constantly changing with the recruitment of new roots and the loss of dead roots. In spring, populations increased as soil temperatures increased (Deans in press), with peak rates of root growth being delayed in the deeper soil horizons. Populations of fine roots decreased as soil dried, the effect being first noticeable in the surface horizons. Subsequently, severe rains, sufficient to wet soil, accelerated root growth in horizons penetrated by the downward movement of the wetting front.

Implications for silviculture in the uplands

Growth rates of forests are influenced by different environmental variables at different stages of growth. After 'crown interlock', growth reflects the balance between (i) increased rates of growth stimulated by days of intense radiation and (ii) the dependence of new growth on current photosynthates and the availability of water. Clearly, sequences of days with strong solar radiation could have a major influence on growth, as could the size of the reservoir of available soil water which can be influenced by silvicultural techniques.

Observations made so far prompt 2 questions about current silviculture practices for Sitka spruce in the uplands of Britain. First, although the depth of ploughing has been decreased and is now more widely-spaced than previously, mainly to encourage lateral root growth and so improve stability (Booth 1974), it may still be detrimental to growth at post 'crown interlock' phases. Savill, Dickson and Wilson (1974) showed that the drainage effect of ploughing at establishment is small compared with the effects of nutrient release from upturned soil, an effect which could be mimicked by the application of fertilizers. Second, current practices encourage very dense canopies at relatively early stages in crop development. In these circumstances, between-tree competition occurs prematurely, detrimentally affecting the assortment of stem sizes and possibly increasing interception losses, thus creating water shortages. On the other hand, when trees are planted densely, they suppress ground vegetation and minimise branching so improving timber quality. Whatever the significance of these effects, however, they are inoperative when trees are 2 m or more in height. Davies (in press) suggested that a proportion of trees at this stage could be 'improved' by decapitation. Alternatively, some effects of close spacing may be replicated with the use of chemical herbicides, by complete cultivation and/or by the use of trees which have been selected for their sparse production of branches, a strongly inherited characteristic.

E.D. Ford, J.D. Deans and R. Milne

References

Aussenac, G. (1970). Action du couvert forestier sur le distribution au sol des précipitations. *Ann. Scis. for.*, **27**, 383–399.

Bingham, J. (1971). Plant breeding: arable crops. In: *Potential crop production*, ed. by P.F. Wareing & J.P. Cooper, 273–294. London: Heinemann.

Booth, T.C. (1974). Silviculture and management of high-risk forests in Great Britain. *Ir. For.*, **31**, 145–153.

Cochrane, L.A. & Ford, E.D. (1978). Growth of a Sitka spruce plantation: Analysis and stochastic description of the development of the branching structure. *J. appl. Ecol.*, **15**, 227–244.

Deans, J.D. (in press). The influence of soil temperature and soil water potential on fine root growth in a 14 year Sitka spruce plantation. *Pl. Soil*.

Davies, E.J.M. (in press). Management implications. In: *The ecology of even-aged plantations*, ed. by E.D. Ford, D.C. Malcolm & J. Atterson. Edinburgh: Edinburgh University Press.

Ford, E.D. & Deans, J.D. (1977). Growth of a Sitka spruce plantation: spatial distribution and seasonal fluctuations of lengths, weights and carbohydrate concentrations of fine roots. *Pl. Soil*, **47**, 463–486.

Ford, E.D. & Deans, J.D. (1978). The effects of canopy structure on stemflow, throughfall and interception loss in a young Sitka spruce plantation. *J. appl. Ecol.*, **15**, 807–819.

Ford, E.D. & Robards, A.W. (1976). Short term variation in tracheid development in the early wood of *Picea sitchensis*. In: *Wood structure in biological and technical research*, ed. by A.J. Bolton, D.M. Catlin & P. Baas, 212–221. Leiden: Leiden University Press.

Ford, E.D., Robards, A.W. & Piney, M.D. (1978). Influence of environmental factors on cell production and differentiation in the early wood of *Picea sitchensis*. *Ann. Bot.*, **42**, 683–692.

Heal, O.W. (in press). Decomposition and nutrient release in even-aged plantations. In: *The ecology of even-aged plantations*, ed. by E.D. Ford, D.C. Malcolm and J. Atterson. Edinburgh: Edinburgh University Press.

Holmes, G.D. (in press). An introduction to forestry in upland Britain. In: *The ecology of even-aged plantations*, ed. by E.D. Ford, D.C. Malcolm & J. Atterson. Edinburgh: Edinburgh University Press.

Milne, R. (in press). Water loss and canopy resistance of a young Sitka spruce plantation. *Boundary-Layer Meteorol*.

Milne, R., Smith, S.K. & Ford, E.D. (1977). An automatic system for measuring shoot length in Sitka spruce and other plant species. *J. appl. Ecol.*, **14**, 523–529.

Savill, P.S., Dickson, D.A. & Wilson, W.T. (1974). Effects of ploughing and drainage on growth and root development of Sitka spruce on deep peat in Northern Ireland. *Proc. int. Symp. Forest Draining*, 241–252. Finland: Society for Forestry.

RESEARCH PRIORITIES FOR SPORTS TURF

(This work was commissioned by the Department of the Environment and largely sub-contracted to the Sports Turf Research Institute—STRI).

Introduction and objectives

A report *Amenity Grasslands—the needs for research* was published by NERC late in 1977. Two considerations prompted the production of the report.

1. Large amounts of valuable energy are expended in creating and maintaining grasslands which are primarily managed for purposes other than production in an agricultural sense.
2. Because amenity grasslands are outside the usual remit of agricultural research, their research requirements may be neglected, particularly in relation to cost-effective management.

Nonetheless, the report inclined to the consensus that mowing and the application of fertilizers were likely to incur appreciable expenditure if many amenity and sports grassland were to fulfil their function. On the more technical aspects, the report suggested that the responses to wear and consequent damage needed to be better understood to facilitate approaches to restoration. For this reason, wear in winter, when recuperative powers of grass are usually minimal and when damage is often severe and likely to persist, was given priority in an experimental programme. In the Netherlands and Germany, seedsmen recommend mixtures with a preponderance of perennial ryegrass *Lolium perenne*, but also containing some smooth-stalked meadow grass *Poa pratensis* and timothy grass *Phleum pratense*, for sites subject to hard wear (Shildrick in press). In contrast, British seedsmen augment *L. perenne* with red fescue *Festuca rubra* and bent grass *Agrostis tenuis* for hockey and football pitches, expecting these species to fill the gaps naturally occurring between plants of the vigorous *L. perenne*. Whereas the recommendations of Dutch and German seedsmen have been validated in experiments testing responses to wear, relatively little is known of the responses of different seed mixtures, and the effects of fertilizers in relation to uncontrollable site factors, including climate. Furthermore, what effect would more realistic wear treatments have on the outcome?

Experimental approach

To answer some of these questions, a collaborative series of factorial experiments, testing the effects of (i) wear and (ii) nitrogen fertilizers on (iii) 16 typical and not-so-typical grass seed mixtures, was established by STRI and ITE (table 1) (Gore *et al.* in press). The near identical experiments, lasting 2 years, were carried out at 3 sites: Bingley in Yorkshire (STRI), Bush near Edinburgh (ITE) and Monks Wood near Huntingdon (ITE). The treatments were arranged in a series of sub-divided plots. Each of the 2 blocks had 2 plots testing with (W_1) and without (W_0) wear. The wear treatment, using a machine specially developed at STRI (cover) (Canaway 1976), consisted of 4 'passes' of this machine each week during (a) January-March 1976 and (b) October-March 1977. Each plot was divided for either 120 kg (N) or 290 kg (N_2) of nitrogen fertilizer per ha per year and each divided plot was further subdivided for 16 grass mixtures (table 1).

During the growing seasons, swards were cut to a height of 25 mm. Although the method has many well-recognized limitations, growth was assessed at 100 point quadrats per subdivided plot, the first leaf hit by the vertical descent of 'the pin' being noted. Assessments were made on 5 occasions after sowing in May/June 1975: (a) at establishment (October/November 1975), (b) post wear (1)—spring 1976, (c) recovery (1)—summer 1976, (d) post wear (2)—spring 1977, and (e) recovery (2)—summer 1977. Percentage cover data were transformed to arcsin $\sqrt{}$ % for statistical analysis.

Results
Differences between sites started to show during germination, i.e. sooner than expected. Whereas the composition of swards at Bingley fairly closely reflected the composition of seed mixtures, *L. perenne* developed at the expense of *P. pratensis* and *A. tenuis* at Bush and Monks Wood (table 2).

By April 1976, after the first round of winter treatment, wear had substantially decreased amounts of cover at Bingley (figure 5). Mixtures containing *Phleum pratense* and *Poa pratensis* (1–7 and 9–11) were usually more resistant than those with *Festuca rubra* and *Agrostis tenuis* (13 and 14), but, in the absence of wear, the latter tended to give more complete cover. Thus, it is possible to appreciate one of the possible reasons for the different mixtures recommended in Britain and the Netherlands, seedsmen in the Netherlands having been influenced by the need to minimize the effects of appreciable wear.

As at Bingley, cover at Bush was substantially less after the first period of wear treatment, but, at Monks Wood (figure 7), there was no significant decrease. Nonetheless, the losses that did occur again reflect the vulnerability of seed mixtures containing large amounts of *Festuca rubra* and *Agrostis tenuis* (figure 6). These results suggest that soil moisture has an important effect on the incidence of wear, the damage during winter 1975/76 being least at the driest site (Monks Wood). Whereas the larger amount of N fertilizer decreased the 'damage' done by wear at Monks Wood, it either had no effect, or tended to increase the losses of cover, at the wetter Bingley and Bush sites.

By the end of the second period of wear treatment (April 1977), the results already discussed were subject to modification. Unlike the 1975/76 winter, rainfall during 1976/77 was similar at all 3 sites. Wear significantly decreased cover at Bingley, Bush and Monks Wood (table 3), the effect still being largest at Bingley, where cover was greatest in the absence of wear and least when subjected to treatment. At this site, the presence of *Festuca rubra* and *Agrostis tenuis*

Table 1.   Mixtures (% of 25 g m^{-2}) of grass seeds sown in 'wear' experiments done at 3 sites

Mixture Number	Lolium perenne	Phleum pratense	Poa pratensis	Festuca rubra	Agrostis tenuis	Comments
1	20	15	40	20	5	
2	20	15	45	20	–	
3	25	20	50	–	5	Mixtures of most species
4	35	28	–	30	7	
5	25	–	50	20	5	
6	–	20	48	25	7	
7	25	20	55	–	–	
9	60	10	30	–	–	Mixtures excluding *Festuca rubra*
10	10	55	35	–	–	and *Agrostis tenuis*
11	10	5	85	–	–	
8	37	–	–	50	12	
16	38	–	–	50	12	Mixtures excluding *Phleum pratense*
12	60	–	–	32	8	and *Poa pratensis*, but including 1%
13	10	–	–	85	5	of *Poa trivialis*
14	10	–	–	55	35	
15	–	10	90	–	–	Excluding *Lolium perenne*, *Festuca rubra* and *Agrostis tenuis*

Table 2. Relation between the mean seed mixture and the resulting stands at 3 sites when assessed 5 months after sowing.

Site	Lolium perenne	Phleum pratense	Poa pratensis	Festuca rubra	Agrostis tenuis
A. Seed mixture, composition of	24	12	34	24	6
B. Sward, composition at					
Bingley	35	23	18	17	7
Bush	67	17	5	10	1
Monks Wood	70	14	3	12	1

Table 3. Effects of a second winter of wear (W_1) on the mean % ground cover of grass swards at three sites (W_0, untreated control swards).

Site	W_0	W_1
Bingley	86	25
Bush	76	40
Monks Wood	78	40

Table 4. Influence of different treatments on the ranges of % cover recorded in experiments done at 3 sites on grass swards of different types with different amounts of nitrogen, and with or without wear.

<————Ranges attributable to————————————>

Stage, in development of experiments	Sites	(a) Mean effects of wear (W_0-W_1)	(b) Effects of nitrogen within the different wear treatments $(W_0(N_2-N_1))$ $(W_1(N_2-N_1))$	(c) Effects of extreme mixtures within the different combinations of wear and nitrogen	Total
Post wear (1) (1976)	BINGLEY	44 / 64%*	6 / 8%	19 / 28%	68 / 100%
	BUSH	39 / 71%	3 / 5%	13 / 24%	55 / 100%
	MONKS WOOD	7 / 33%	6 / 26%	9 / 41%	23 / 100%
Post wear (2) (1977)	BINGLEY	37 / 71%	1 / 2%	14 / 27%	52 / 100%
	BUSH	20 / 61%	3 / 9%	10 / 30%	34 / 100%
	MONKS WOOD	23 / 56%	1 / 3%	17 / 41%	42 / 100%
	MEAN (excluding Monks Wood 1976)	33 / 66%	3 / 6%	15 / 29%	50 / 100%

*Range of specific treatment as % of total range of cover.

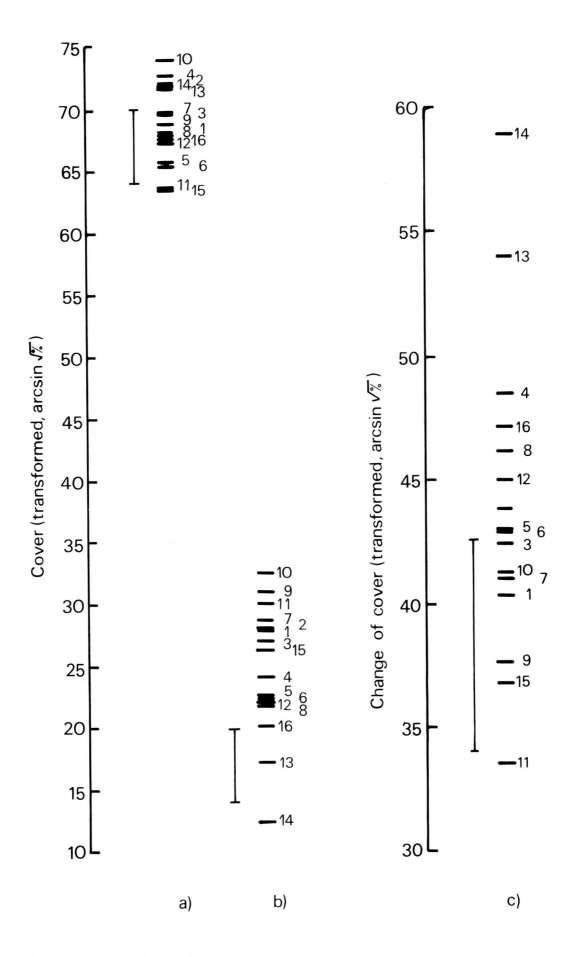

Figure 5   Cover provided by 16 grass mixtures at Bingley, Yorkshire, when assessed in April 1976 after being subjected to wear during winter 1975/76. (a) the untreated control series (W_0), (b) the series subject to wear (W_1), and (c) the amounts of cover lost by wear (W_0-W_1). Vertical bars show least significant differences at 0.05 level.

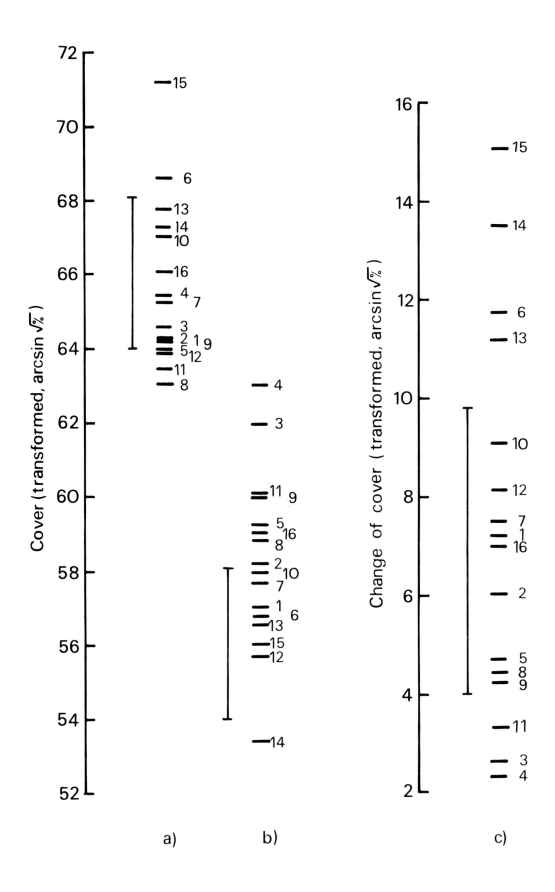

Figure 6   Cover provided by 16 grass mixtures at Monks Wood when assessed in April 1976 after being subjected to wear during winter 1975/76. (a) the untreated control series (W_0), (b) the series subject to wear (W_1), and (c) the amounts of cover lost by wear (W_0-W_1).

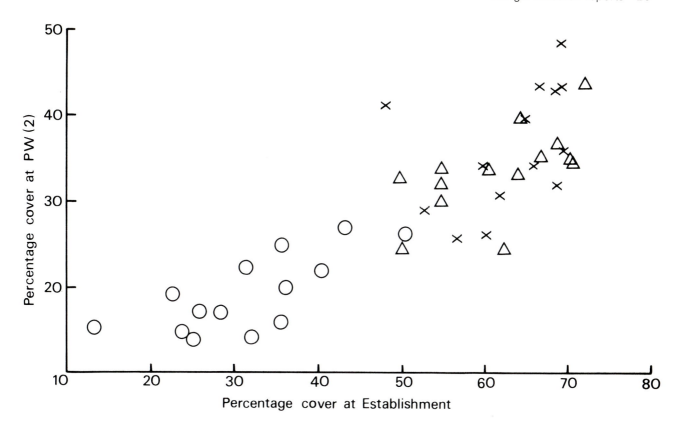

Figure 7 Relationship between % cover (1) provided at establishment by the Lolium perenne component of 14 grass mixtures and (2) remaining after wear treatments applied during the winters of 1975/76 and treated 1976/77. Data from three sites: O, Bingley; △, Bush and X, Monks Wood.

may be advantageous in the absence of wear, but a liability when wear occurs. Interestingly, detailed examination of the data, collected after the second winter of wear, indicated that the cover attributable to *L. perenne* was directly related to the abundance of this species at establishment, 18 months previously (figure 7). This observation suggests that other species in the differing mixtures had little effect on the development of *L. perenne*.

Discussion
The results of the collaborative series of experiments suggest that mixtures containing *Lolium perenne*, *Phleum pratense* and *Poa pratensis* are appropriate for sites consistently subject to severe wear. However, where wear is less intense, the British practice of including *Festuca rubra* and *Agrostis tenuis* is understandable, always remembering that amounts of cover given by the different mixtures did not differ greatly in absolute terms.

In developing a research programme, it is essential to order priorities correctly. By estimating the effects attributable to different factors (table 4), it would appear that wear treatments had larger effects on percentage cover than nitrogen fertilisers and the different seed mixtures 60–70 per cent v 3 per cent v 30 per cent respectively. Additionally, there is evidence to support the idea that wear damage is greater in wet than in dry conditions. Thus, wear caused a decrease of

only 7·4 per cent at Monks Wood during the 'dry' winter of 1975/76 in contrast to 20·4–43·9 per cent losses in other years and at other sites. This evidence suggests that the stability of grass swards subjected to wear might be maintained if the build-up of soil moisture were minimized. Although interactions have been neglected in this summary, some were nevertheless detected. Wear damage was usually greater in plots with 290 kg N ha^{-1}yr^{-1} than in plots with 120 kg N ha^{-1}yr^{-1}, although the reverse was suggested at Monks Wood after the dry 1975/76 winter.

Although it would be wrong to be too definite on the basis of results of a single set of experiments, it seems likely that the scope for improving performance by modifying recommendations about the choice of species in seed mixtures is rather limited, and such a conclusion would have implications for priorities for future research.

A.J.P. Gore and R. Cox

References
Canaway, P.M. (1976). A differential slip wear machine DS1 for the artificial simulation of turfgrass wear. *J. Sports Turf Res. Inst.*, **52**, 92–9.
Gore, A.J.P., Cox, R. & Davies, T.M. (In press). Wear tolerance in turfgrass mixtures. *J. Sports Turf Res. Inst.*
Shildrick, J. (in press). Turfgrass mixtures in the UK. *Proc. int. Turfgrass Research Conf. 3rd, Munich.* (American Society of Agronomy, Madison, Wisconsin, USA.)

ECOLOGY OF A WOODLAND TOADSTOOL, *MYCENA GALOPUS*

Many fungi have an important role to play in the decomposition of plant litter and the recycling of nutrients in forest ecosystems. One of these, *Mycena galopus*, which occurs in temperate regions, forms toadstools with a small brown cap topping a stalk (stipe) c. 5 cm tall, the stipe containing a white milk-like substance. Less conspicuously, *M. galopus* produces superficial white hyphal strands and mycelial wefts with hidden thread-like hyphae penetrating dead plant tissues. *M. galopus* is a saprophyte belonging to the Basidiomycetes (Plate 16, cover).

Interest in *M. galopus* dates back to the International Biological Programme (IBP) which highlighted gaps in our knowledge of the ecology of fungi. At the start of a project at Meathop Wood on the Cumbrian coast, it was impossible to give a name to the key decomposers and the biomass of soil fungi was unknown: weights and numbers of toadstools in a nearby area, collected over 3 seasons, provided the only available data (Hering 1966). Here, *M. galopus* represented 31 per cent of the total number, but only 3 per cent of the total weight of fruit bodies.

The history of microbiology has been called a history of techniques. This is particularly true of fungal ecology, where research has been severely constrained by the availability of techniques for use in the field. Until recently, it was only feasible to estimate the biomass of fungal mycelium in litter and soil with the agar-film technique by which lengths of hyphae in agar films, containing known amounts of dispersed substrate, are measured microscopically, biomass being subsequently calculated assuming that hyphae are cylindrical. Although this method underestimates fungal biomass, it nevertheless gives estimates that are biologically reasonable (Frankland, Lindley and Swift 1978).

The total fungal biomass (live and dead) in Meathop Wood from surface litter down to bedrock, excluding that in timber, was c. 300 kg ha^{-1}, equivalent to 4 per cent of the total bacterial biomass (T. R. G. Gray, unpublished data). When the living fraction of the biomass was separately estimated, by assessing the volume of cytoplasm in mycelium and assuming that cell contents were indicative of life (Frankland 1975a), it was found to amount to 74 kg ha^{-1}, to a soil depth of 33 cm, with 22 per cent of the mycelium occurring in the thin covering layer of leaf litter where its weight was 8 times heavier than that of living bacteria. This distribution tallies with that inferred from measurements of respiration (Spink 1975).

Litter is colonized by a diverse array of fungi. Many were isolated and identified from mycelium growing from fragments of litter which were previously washed and put on agar media, and others were isolated from spores. In this way, species lists could have been compiled, but the Basidiomycetes—causing extensive white rot when breaking down the principal structural components of litter—were rarely represented unless litter was incubated in moist conditions for 2–3 months, when fruit bodies of Basidiomycetes appeared. The latter were often immature or malformed, but characteristic features, such as the white milk of *M. galopus*, facilitated identification. *M. galopus* was found, 2 years after leaf-fall, on more than 80 per cent of the leaves of oak, the dominant tree at Meathop. It was also found colonising the litter of ash, hazel and birch following earlier colonisation by primary saprophytes. Thus, it seemed that *M. galopus* was a key decomposer, but how could its role be quantified, remembering that its hyphae lack distinguishing morphological features?

To assess its ability to (a) decompose different types of litter and (b) capture plant nutrients, it was necessary to employ pure culture techniques using (i) litter sterilised by gamma-radiation with the help of the Atomic Energy Authority and (ii) isolates of *M. galopus* cultured from cap tissues (of fruit bodies), spores and mycelial strands. The use of gamma-radiation minimized problems associated with the troublesome (a) survival of bacteria and (b) release of soluble carbohydrates occurring when some other sterilants are used (Howard and Frankland 1974). For completeness, it was still necessary to quantify the production of mycelium. After incubating for 6 months, mycelial wefts were laboriously dissected from litter with fine forceps and chemically analysed, whereas hyphae within macerated litter were measured *in situ*, amounts of hyphae and mycelium being subsequently corrected to give the standing crop biomass which is less than the total production of mycelium during the course of the 6-month experimental period.

Total production = net change in biomass + loss of biomass (by decay or consumption).

To estimate rates of mycelial decay, glass slides were put in close contact with litter. Subsequently, the extent of wall breakdown (autolysis) in mycelial overgrowths was assessed, using a phase-contrast microscope. In the event, hyphal wall breakdown was not detected, so the estimates of biomass based on hyphal length were equated with production by assuming that conditions on glass reflected those in litter. The use of phase-contrast, by pinpointing empty hyphae which were previously invisible, led to increased estimates of the biomass (Frankland 1974). It was confirmed that *M. galopus* could be an active decomposer. When incubated on hazel litter for 6 months, it decomposed about a quarter of the lignin and cellulose, a third of the soluble carbohydrates, some crude protein and tannins, and most of the crude fat. On other litters common in Meathop Wood, it was 4–17 per cent 'efficient' as judged by the formation of mycelium. In forming mycelium, it utilised, rather than released, plant nutrients, capturing 3–7 per cent N, 5–15 per cent P

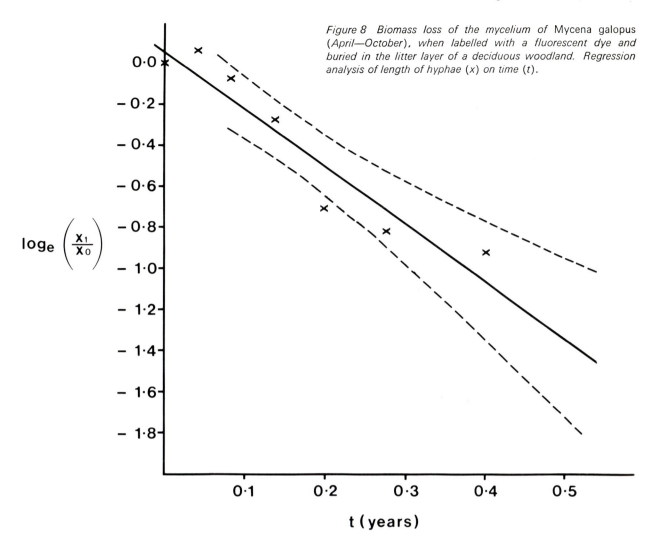

Figure 8 Biomass loss of the mycelium of Mycena galopus (April—October), when labelled with a fluorescent dye and buried in the litter layer of a deciduous woodland. Regression analysis of length of hyphae (x) on time (t).

and 2–6 per cent K. To link these laboratory observations to events in woodland, it was necessary to obtain estimates of the annual production of M. galopus in woodlands. This estimation was to prove relatively simple by studying the hyphae of M. galopus which can easily be recognised in oak litter, a substrate usually dominated by M. galopus alone.

Biomass (standing-crop) estimates were calculated from measurements of hyphal length using the agar-film technique, but, in order to determine the loss of mycelium through decay or grazing, a new method had to be devised (Frankland 1975b). To follow its fate, mycelium of M. galopus was labelled with Calcofluor White, a fluorescent brightener. Discs of oak litter colonised by labelled mycelium were buried in litter, and lengths of surviving fluorescing hyphae assessed periodically. The dye was neither translocated to new growth nor leached by rain: furthermore, it did not appear to affect the palatability of mycelium to small grazing animals. From regression analyses, it was calculated that 0·5 kg ha^{-1} of M. galopus mycelium was produced in 1 year in decaying leaves, an amount 10 times greater than that for the production of fruit bodies. This production suggests that less than 6 per cent of the N, P, K of oak leaves was incorporated into

fungal tissue, a possibly significant amount bearing in mind that M. galopus is only one of many colonizing organisms (figure 8).

With these different pieces of information, it is now possible to be within an order of magnitude in assessing the activities of M. galopus in Meathop Wood, but, for these estimates to be given greater precision, it is necessary to develop more precise methods of locating mycelium. To this end, the fluorescent-antibody (FA) technique has been developed with Prof. T. R. G. Gray at Essex University. Antigen—antibody reactions located with fluorescent compounds can provide a highly sensitive means of detecting specific microbes, but leaf litter poses many problems. Using rabbits, antisera against M. galopus have already been produced and linked (conjugated) with the dye fluorescein isothiocyanate. At first, the antisera were not specific to M. galopus, and they cross-reacted with other fungi found in oak litter, but these reactions have now been eliminated or minimized. Lengths of FA-reactive M. galopus hyphae were estimated using a grid-intersection method, and these lengths were subsequently converted to fungal biomass as in the agar-film method.

If the FA-reaction discriminates between living and

dead mycelium, it might be used specifically to estimate viable biomass of *M. galopus*, so aiding our ecological understanding of this fungus in many different types of woodland, both deciduous and coniferous.

J.C. Frankland, A.D. Bailey and P.L. Costeloe

References
Frankland, J.C. (1974). Importance of phase-contrast microscopy for estimation of total fungal biomass by the agar-film technique. *Soil Biol. Biochem.*, **6**, 408–410.
Frankland, J.C. (1975a). Estimation of live fungal biomass. *Soil Biol. Biochem.*, **7**, 339–340.
Frankland, J.C. (1975b). Fungal decomposition of leaf litter in a deciduous woodland. In: *Biodegradation et Humification*, ed. by G. Kilbertus *et al.* 33–40. Sarreguemines: Pierron.
Frankland, J.C. & Collins, V.G. (1974). A bacterium in *Quercus* leaf litter resistant to sterilizing doses of gamma-radiation. *Soil Biol. Biochem.*, **6**, 125–126.
Frankland, J.C., Lindley, D.K. & Swift, M.J. (1978). A comparison of two methods for the estimation of mycelial biomass in leaf litter. *Soil Biol. Biochem.*, **10**, 323–333.
Hering, T.F. (1966). Fungal decomposition of oak leaf litter. *Trans. Brit. Mycol. Soc.*, **50**, 267–273.
Howard, P.J.A. & Frankland, J.C. (1974). Effects of certain full and partial sterilization treatments on leaf litter. *Soil Biol. Biochem.*, **6**, 117–123.
Spink, M. (1975). *Growth and metabolism of soil fungi*. PhD. thesis, Liverpool Polytechnic.

THE INVERTEBRATE FAUNA OF DUNE AND MACHAIR SITES IN SCOTLAND

(This work was commissioned by the Nature Conservancy Council as part of its programme of research into nature conservation)

Introduction

This work was commissioned by the Nature Conservancy Council to assess the scientific interest of the invertebrate fauna associated with the same range of dune systems studied by the botanists from 1975 to 1977 (see ITE Annual Report 1976, p. 39 for map of sites). The botanical survey covered 94 sites on the east, north and west coasts, including the Inner and Outer Hebrides, Orkney and Shetland. It was not possible to include all these in the invertebrate survey and the total was reduced to 58 sampling areas. These included 4 large sites where 2 trapping stations were established on each. Most of the fieldwork for the invertebrate study was done during June and July 1976, but 11 sites (in the Moray Firth area) could not be sampled until June and July 1977. The 1976 season coincided with a record drought period in Britain, but this was not experienced throughout the whole of Scotland. For example, the survey teams who operated simultaneously on the east and north coasts and in the Outer Hebrides experienced a marked difference in the weather conditions from east to west. It was a good deal wetter and windier on the west coast, and in the Hebrides, than on the north or east coasts. On the other hand, the weather conditions in the Moray Firth area in 1977 were average for this part of Scotland (Plates 1, 3, 5).

Objectives

The objectives as defined in the project plan are:

1. To provide information which, together with data from the botanical survey, will form a basis for conservation evaluation of individual sites, or groups of sites, on sand dune formations and machair along the east, north and west coasts of Scotland, including the Outer Hebrides.

2. To extend the existing knowledge of the occurrence and distribution of Lepidoptera, Coleoptera, Araneae, terrestrial Isopoda, Mollusca (land snails), and the Diplopoda.

3. To assess the validity and practicability of extensive "quick" survey for groups of widely separated sites in logistically difficult situations.

Methodology and siting of traps

The simultaneous sampling of the invertebrate fauna at 47 sites, as was done in 1976, posed problems of standardising the sampling procedure and also of ensuring that the selection of sites in the field and the operation of traps were the same in all cases. Sand dunes and machair have the advantage of being relatively simple systems, although the degree of habitat complexity was found to be considerably greater than was at first thought likely. In order to achieve simultaneous sampling, 3 teams were formed with responsibility to survey, respectively, the Outer Hebrides and regions designated as North Coast and East Coast. In 1977, a fourth party sampled the 11 sites in the Moray Firth area.

The most effective trapping method for ground-living invertebrates, particularly those which are active, such as Araneae, Coleoptera, Diplopoda and Isopoda, is to use pitfall traps which operate continuously and produce the largest amount of material. Hand-collecting would probably have been more effective for sampling the land snails, but this method could not be included in the programme for lack of time. The Lepidoptera were collected by using a battery-operated light trap which switched on and off automatically at dusk and dawn.

Because considerable quantities of equipment had to be taken into the field, the pitfall traps and the light trap with battery had to be grouped within a fairly confined area. This requirement limited the range of habitat diversity available for sampling on each site. The 4 pairs of pitfall traps were arranged at the corners of a square, or else along a transect, with the light trap in the middle. A further constraint was the need to position the light trap out of sight of the general public and where it was not likely to be a hazard to shipping. In addition, pitfall traps on dune systems must be sited where there

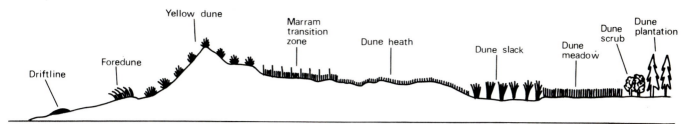

Figure 9   A diagrammatic representation of animal habitats on a dune system showing the position of the marram transition zone, (Duffey, 1968).

is sufficient vegetation cover to shelter them from the action of the wind because of sand-blow. On most of the Hebridean sites, the dune system was not well-formed topographically and the traps were positioned in very short grazed turf, where marram grass was a minor constituent of the flora (see plates 2, 4, 7). Elsewhere on the Scottish coast, the traps were generally situated in what has been described as the marram transition zone (figure 9). This zone is situated on the landward side of the main dune ridge and is characterized by vegetation in which marram grass is no longer dominant, where bare sand is localised and where there is often a very varied flora.

Trapping period and number of collections
In both 1976 and 1977, the pitfall traps were put in place during the first or second week of June and operated continuously for about 6 weeks. They were emptied after the first 7–8 days, at the end of the following 3–4 weeks and finally after the last 7–8 days. The light traps were operated for only the first and last weeks of the total period, because the batteries had to be recharged after 7–8 days. At the end of the fieldwork programme, there were 116 light trap catches and 1,392 pitfall trap catches to be sorted and the animals identified.

Methods of analysis
The sampling methods used during the survey were selected to achieve uniformity of catch so that comparisons could be made between sites, between groups of sites and between sampling periods. However, the light traps were subject to unforeseen mechanical faults which were difficult to quantify and there were climatic and habitat differences between the 4 geographical areas, which also influenced the catches. The pitfall traps were less subject to malfunction or to variation of catch due to weather. The recommended statistical treatment was ordination by reciprocal averaging followed by an indicator species analysis (Hill 1973). This analysis was followed by the calculation of diversity indices for each site and a single factor analysis of variance for each of the 4 geographical regions.

The fauna and its range of variation
The total recorded fauna included 656 species as follows: Coleoptera-Carabidae 55, Coleoptera-Hydrophilidae to Scolytidae 289, Lepidoptera 158, Araneae 115, Mollusca (land snails) 26, Diplopoda 9 and terrestrial Isopoda 4. The ordination was based on the pooled data for each site and calculated for each major invertebrate group, region by region. The distribution of points on the ordination graphs shows that the Carabidae form the most compact groups in each of the 4 regions and also have the greatest degree of overlap. This suggests that the variation in the carabid fauna from region to region was less than that recorded for other invertebrate groups. The ordination graph for the other families of Coleoptera, which totalled 289 species, shows 4 well-separated groups with comparatively little overlap. The North Coast region shows the greatest spread, while there is least variation in the Outer Hebrides. The spider fauna shows a clear separation between the 4 regions, although there is considerable overlap between the North Coast and the East Coast. The Lepidoptera analysis is based only on the July sample because a large proportion of light traps did not function correctly in the first trapping period. The most compact group is formed by the North Coast sites, and there is some overlap with the East Coast. The Hebridean group is slightly more widely spread, and the Moray Firth formed a distinct group, well-separated from the other three regions. The scatter of the points on the ordination diagram clearly showed that geographical differences were important factors separating the regions. However, a number of individual sites were either isolated or associated with a different geographical area and in most cases habitat differences could be identified as the probable cause.

The single factor analysis of variance for (a) total species, and (b) total specimens in all groups confirms the general impression that the East Coast sites showed the greatest richness of species, but the Hebridean sites produced the largest numbers of individual animals. Significance levels of differences between the regions for each major invertebrate group also confirmed the distinctiveness of the East Coast in relation to the others. A similar analysis using diversity indices (Shannon's and Simpson's) did not reveal any further information on the relationships between the regions.

A comparison by eye of the arithmetic differences between the number of species and number of individuals for the fauna from region to region bears out the results of the ordination analysis. The mean total of species increases in sequence from the Outer Hebrides through the North Coast and Moray Firth to the East Coast—75, 86, 94 and 135. Some of the major groups also follow this trend quite closely, for example the Lepidoptera and Araneae.

	O. Hebrides	N. Coast	Moray Firth	E. Coast
Lepidoptera	9	22	35	38
Araneae	12	18	20	24

On the other hand, the Carabidae (11, 7, 9 and 12) and the other families of Coleoptera (36, 31, 26 and 49) do not follow this trend. In contrast, the land snails show a reverse trend with more species on the calcareous shell-sand sites in the Outer Hebrides—6, 5, 2, 4. The total numbers of snails show the same pattern (1121, 156, 7, 46), and, similarly, the total numbers of Lepidoptera for each region (131, 250, 423, 613) support the species trend, but this result was not repeated for the Carabidae and other families of Coleoptera, nor for the spiders.

The influence of habitat differences
The sequential increase in mean numbers of species of Lepidoptera and Araneae from the Hebrides to the East Coast seems to be reflected in habitat differences, possibly modified by climatic characteristics. The greater range of habitat types on the East Coast included heathland, scrub and plantations. For example, at Tentsmuir the large number of moths trapped included 9 species not recorded elsewhere and mostly characteristic of woodland.

Tolsta was the only Hebridean dune system situated on the East Coast (Isle of Lewis) and also where marram grass formed the dominant vegetation instead of a machair turf. In the ordinations and indicator species analysis, this site consistently falls within the North Coast regional group. On the other hand, the short dense turf of Sheilgra, on the mainland west coast, produced a fauna which was closer, in the ordinations and ISA, to that of the Outer Hebrides. Lossiemouth is totally isolated, in all Coleoptera ordinations except one, from the rest of the Moray Firth group and also from all other sites. Its beetle fauna was not rich, but it scored the largest numbers of *Philopedon plagiatus*, *Otiorrhynchus atroapterus* and *Orthocerus clavicornis*. It has been suggested that the presence of these characteristic coastal dune species reflects the earlier extensive open dune formations which were present prior to afforestation. Gullane (East Coast) showed no unusual features in the analyses, but provided the most remarkable catch of the survey. A single male of the spider *Erigone aletris*, a North American species pre-viously unrecorded in Europe, was trapped. Several more males together with females were taken by R. Snazell in 1978.

One of the most valuable results of the survey has been the considerable extension of knowledge on the distribution and ecology of many invertebrate species. For example, *Lycia zonaria* (the belted beauty moth) was taken, as larvae, on 16 out of 18 sites in the Outer Hebrides. Although previously known from the Outer Hebrides, elsewhere in Britain it is scarce and mainly confined to the coasts of north Wales and Lancashire. Similarly, the abundance of the spider *Erigone promiscua* in the Outer Hebrides and north-west Scotland was previously unknown, while the lycosid *Pardosa purbeckensis*, which is confined to the intertidal zone of English coasts, was frequently trapped on the machair dune turf in the Outer Hebrides (Plate 6).

E. Duffey

Reference
Hill, M.O. (1973). Reciprocal averaging: an eigenvector method of ordination. *J. Ecol.*, **61**, 237–249.

RESEARCH ON OYSTERCATCHERS AND MUSSELS ON THE EXE ESTUARY

Introduction
Research began in 1976 to study the effect of oyster-catchers (plate) on the numbers of shellfish which provide most of their winter food. There are two main reasons for this research. First, it provides an opportunity for studying the interaction between a population of predators and its prey, and so contributes to our knowledge of a fundamental ecological process. Second, there are two applied problems which we wanted to study. One stems from the frequent claim that oyster-catchers are a pest of commercial shellfisheries, and it is intended to study this claim in more detail than has been possible previously. The other arises from the increasing pressure put on the waders' winter feeding grounds from schemes developing them for agricultural, recreational and industrial purposes. Such schemes raise the question as to whether the feeding areas are already fully utilised or whether spare capacity exists for birds forced by these developments from their present feeding grounds. Our aim is to model the interaction between oystercatchers and shellfish in order to simulate the effect of a sudden increase or decrease in the numbers of birds on the long-term interaction between them. Partly this research will help us to understand the way the system works, but it will also provide a better basis for predicting what might happen if oystercatcher numbers were to be reduced in an attempt to alleviate their impact on shellfish or, conversely, to be increased following a loss of feeding grounds elsewhere.

One reason for choosing the oystercatcher for study was that it is the species of wader most frequently accused of being a pest and, indeed, several thousand were shot as a consequence in south Wales some years ago. In addition, it is particularly suitable bird for the kind of work involved. It is large and easily counted, its age can be estimated easily, and its feeding and social behaviour are conveniently measured. The Exe estuary was chosen because it is accessible, of reasonable size (figure 10), and is relatively isolated from other estuaries

Numbers of oystercatchers and their food on the Exe
Oystercatchers are counted at low water on the estuary itself, along the adjacent coast, and in the fields surrounding the estuary. The numbers recorded during the first year of the study are shown in figure 11. Numbers. rise sharply in late summer as the adults return from their breeding areas in northern Britain, Norway and Holland. Some birds leave in October, but most remain until February when adults begin their migration back to the breeding grounds. Only non-breeding immature birds are left from April onwards, and little change in numbers occurs until the adults again return. Ringing has shown that most individuals remain on the Exe from autumn to spring and that most adults return there each year.

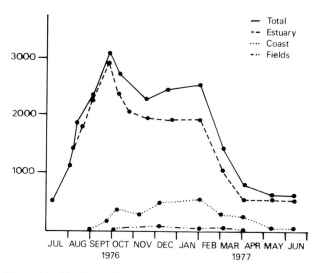

Figure 11 Numbers of oystercatchers on the Exe estuary, adjacent coast and fields on the counts made at low water.

Figure 11 also indicates the numbers of birds feeding in the 3 major habitats. Most oystercatchers feed on the estuary itself. Some, particularly the younger birds, forage on the mud and sand flats and eat cockles *Cerastoderma edule*, ragworms *Nereis diversicolor*, or a deep-burying shellfish *Scrobicularia plana*. Most, however, eat mussels and the majority of adults are found on the 31 mussel beds situated in the lower half of the estuary.

A few also feed in the fields at low water, and most of these are birds in their first winter, eating earthworms and leatherjackets. More birds of all ages feed in the fields at high water, but mainly in winter to supplement the reduced amount of food they are able to find then on the estuary at low water. Others feed on the coast at low water, particularly in winter when disturbance from people is least. Birds of all ages feed there, but especially first-winter birds. They eat the common mussel *Mytilus edulis*, and a sand-dwelling bivalve *Spisula*.

The main food of the population is undoubtedly the common mussel and between 1500 and 2000 birds feed almost entirely on this species between August and March. Oystercatchers eat mussels by prising, or breaking, open the shells of animals pulled from the beds. Examination of shells left by the birds reveals that they mainly eat the large ones, particularly those between 45 and 65 mm long (figure 12). Because of this, and since rather few mussels are taken by the young birds during the summer, the project resolves itself into a study of mainly winter predation by oystercatchers on the older individuals within the mussel population.

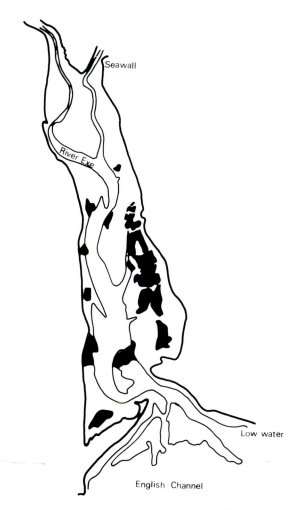

Figure 10 Exe estuary, and adjacent coast. Black areas show approximate locations of the main mussel beds.

Life history of mussels
In the summer, each female mussel produces some thousands of eggs which are fertilised in the sea. After a few weeks in the plankton, the larva grows a shell and sinks to the bottom where it may try several substrates

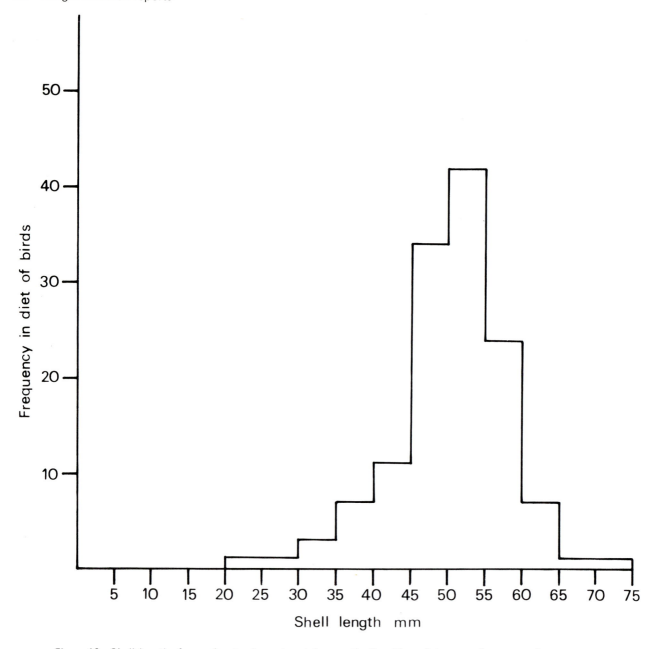

Figure 12   Shell-length of mussels eaten by oystercatchers on the Exe. Most of the mussels eaten are large ones.

before finally settling. Delaying final settlement in this way enables many larvae to be too large to be eaten by the adult mussels. The survival and growth rate of young mussels varies from place to place, depending on the density and size of their neighbours, among other factors such as the period of tidal immersion and so the amount of time available for feeding. Populations of mussels in estuaries often have low recruitment but good growth rates, with a relatively high proportion surviving to 5 or 10 years of age. This high rate of survival contrasts with that of populations on rocky shores. At low levels on the beach, recruitment may be large and growth rate very fast, but few mussels survive beyond 1 or 2 years because of heavy mortality by other predators from the sublittoral. In contrast, at high levels on exposed rocks, recruitment is low and growth is poor, but, in the absence of predators, many survive for a long time. These findings on rocky shores suggest that predators can have a great effect on the performance of mussel populations, so that a study of the interactions between an important predator on estuaries, the oyster-catcher, and its mussel prey should be worthwhile.

Interactions between oystercatchers and mussels
There are likely to be 2 main responses shown by the birds to changes in the abundance of the preferred size class of mussels. First, more birds might eat these mussels as their abundance increases, either because more birds come to the estuary, or because a greater proportion of those already there take this prey. Second, as the abundance of the mussels increases, individuals might eat more per day until all their food consists of mussels. Both responses in combination determine the total numbers of mussels eaten. This predation on the mussels will then affect the settlement and growth

Plate 10   Observation tower and the four large enclosures used for studying rabbit behaviour.   *Photo: D T Davies*

Plate 11   Within the 2-acre paddock, small enclosures of various sizes have been erected to investigate the effect of living space and density on the reproductive potential of the wild rabbit.   *Photo: D T Davies*

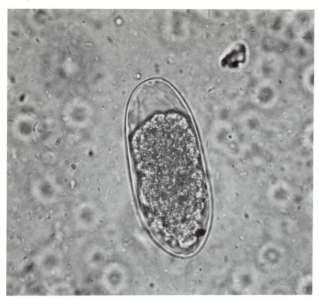

Plate 12   The egg of the caecal threadworm Trichostrongylus
tenuis *of red grouse.*
Photo: G R Wilson

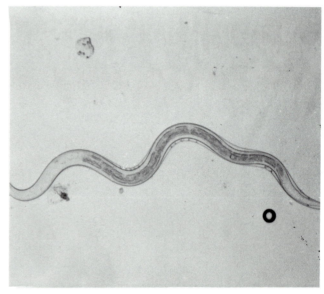

Plate 13   The third larval stage of T. tenuis *showing the sheath
which protects it while it waits to be eaten and infect grouse.*
Photo: G R Wilson

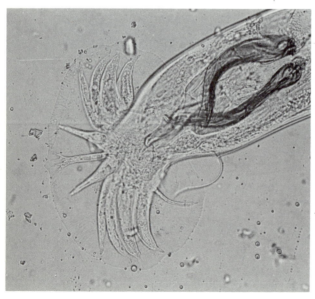

*Plate 14   The characteristic spicules and copulatory bursa of a
male* T. tenuis.
Photo: G R Wilson

potential of the smaller animals on the beds and the reproductive output of the population as a whole. These prey responses in turn will, in part, determine the future food supplies of the birds, and hence their impact on the prey in subsequent generations. Because factors other than the abundance of mussels affect the feeding activities of the oystercatchers and, likewise, factors other than the birds affect the performance of the mussel population, other variables must also be incorporated into both models if a proper understanding is to be obtained.

These simplified ideas are summarised schematically in figure 13. Our aim is to construct 2 models, one of the mussel population and one of the bird predation on it. The output from one model will, of course, constitute the input for the other. Much of the fieldwork on the Exe is concerned with describing the feeding responses of the birds to variations in mussel abundance and with measuring changes in the abundance of the mussels themselves. Much remains to be done, so all this report can do is to present some of the preliminary findings.

Studies on mussels
The mussel beds are sampled twice a year to establish the numbers of each age and size class in each bed. The surveys are carried out in September, when most birds have arrived and mussel growth and reproduction are declining prior to the winter, and in March when the birds are leaving and the mussels again begin their summer growth. Each bed is mapped and the proportion covered with mussels is measured. Random core samples are taken to estimate the abundance of all age and size classes present. The procedure gives estimates of mussel density for individual mussel beds

with standard errors varying between 5 and 35 per cent of the mean, and for the population on the whole estuary of between 5 and 15 per cent. Given the large variations in mussel abundance that are expected, these estimates are sufficiently precise for our purposes.

Data from the first winter of the study (table 1) show that the highest rates of mussel loss occur in the larger size classes taken by oystercatchers. Estimates of the daily consumption of mussels (see below) by 1500–2000 oystercatchers during the winter indicate that most of this loss can be attributed to the birds themselves.

Eventually, the data from mussel surveys will be used to construct a life table of the whole mussel population, but, after only 2 years, this construction cannot yet be attempted. However, the early results reveal some interesting trends emerging from comparisons made between mussel beds at one time. For instance, more young mussels settle on beds where larger mussels are already abundant (figure 14). Furthermore, a greater

Table 1. The disappearance of mussels of various sizes between September 1976 and March 1977. The large ones (over 45 mm) eaten by oystercatchers have the highest rate of loss.

Size (mm)	Total nos in (x1000)		Percentage Decline
	September	March	
1–10	60713	56456	7
11–20	56422	50175	11
21–30	81934	71862	12
31–45	158687	148121	7
46–60	118145	85997	27
61+	4543	2915	36

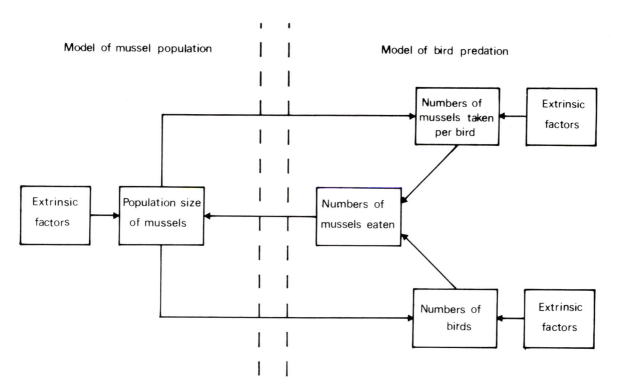

Figure 13 Simplified diagram of the predatory-prey interaction between oystercatchers and mussels.

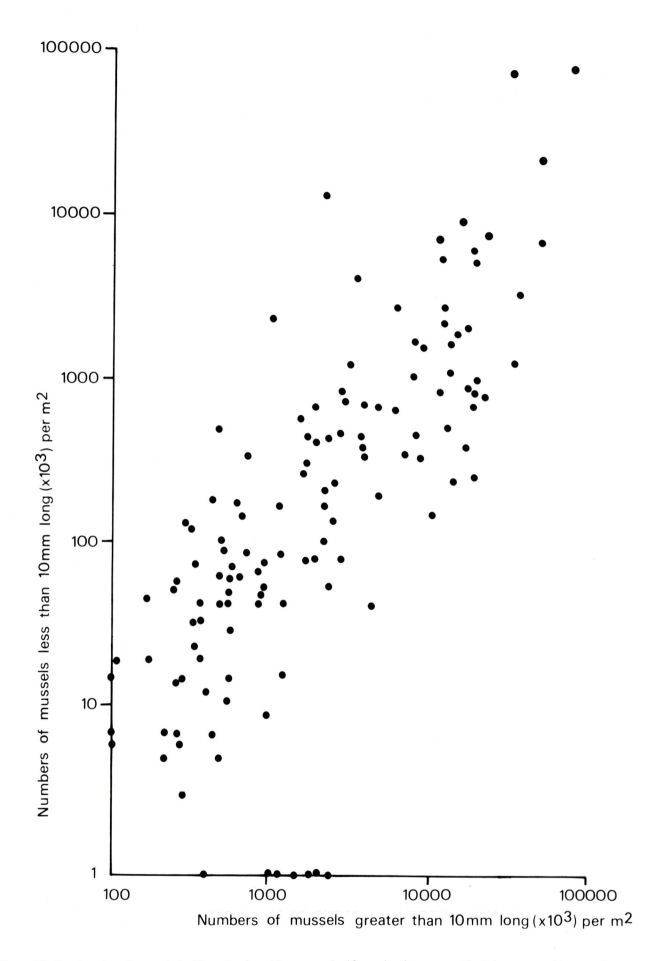

Figure 14 Density of small mussels (<10mm long) and large ones (>10 mm long) on mussel beds in autumn. More small ones settle where larger ones are abundant than where they are scarce.

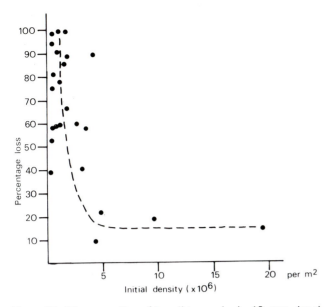

Figure 15 The mortality of small mussels (<10 mm long) between autumn and spring in relation to their density in the autumn. More disappear where few settled at the start.

figure 16 for the two most important factors, mussel density and distance to roost. When few birds are present, bird density only correlated with distance to roost: in other words, most birds are found on the few beds near to the roost. However, as total numbers rise, the birds spread out over the other beds, apparently because there is a limit to the numbers of birds that can feed on the beds near to the roost (figure 17). As they spread out, the density of birds becomes increasingly closely correlated with the density of mussels. Indeed, mussel density eventually becomes more important than distance to roost in accounting for bird density.

Although tests of the causality of these relationships are being planned and other factors which could influence bird density are being measured, it is still worthwhile pointing out the possible importance of these findings. First, they suggest that there is a maximum density of birds that can feed on the beds, presumably set by the behaviour of the birds themselves towards each other. This density is already reached on some beds (figure 17), but not on others (figure 17). There-

proportion of those that settle disappear during winter from those beds where large mussels are scarce (figure 15). In other words, more young mussels settle where mussels are already abundant and, once there, they also survive better. Since oystercatchers eat only the larger mussels, it is important to find out how their predation might affect the settlement and subsequent survival of young ones, and field experiments are in progress to study this effect.

Responses of oystercatchers to mussel abundance
It is too soon to see if the total number of oystercatchers which feed on mussels varies from year to year according to the abundance of the shellfish on the estuary. It is, however, already possible to examine how the abundance of mussels affects the numbers of birds feeding on the different mussel beds within one winter.

In addition to the density of the mussels themselves, several features of the mussel beds which might affect the numbers of birds feeding on each were measured during the winter. Multiple regression analysis suggests that 3 factors may be important in influencing bird density on a mussel bed. The density of mussels and their size are both important, with birds not surprisingly preferring beds which have high densities of large mussels. In addition, the distance of the bed from the roost used by the birds at high water is important, with the highest bird densities being recorded on mussel beds near the roost, presumably because they need spend least time and energy in flying to them.

An important finding is that the relative importance of these factors varies with the total numbers of birds feeding on all the mussel beds. This is illustrated in

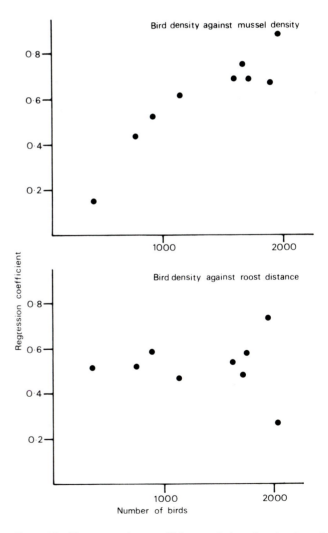

Figure 16 The regression coefficients relating the density of oystercatchers on different mussel beds to the density of mussels and the distance of the bed from the roost. The importance of mussel density increases as the total numbers of birds feeding on mussels on the whole estuary increases, whereas the distance of a bed from the roost is always important.

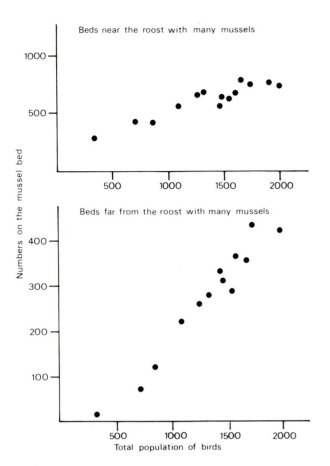

Figure 17   The numbers of oystercatchers on dense mussel beds (a) near the roost, and (b) far from the roost in relation to the total numbers of birds feeding on mussels. The numbers on the mussel beds in (a) tend to level off when large numbers are present on the estuary, whereas there is no indication of this in (b).

fore, any increase in bird numbers on the mussel beds as a whole is likely to occur on some beds only, namely those that are far from the roost and have low densities of mussels. This relationship must be recognised if the simulation experiments with the model are to be at all realistic. Second, the converse also applies: any reduction in bird numbers is likely to result in most of the survivors feeding on the dense beds near to the roost. Since the feeding rate of oystercatchers varies from bed to bed, their predation over the whole estuary will change in a complicated way as the oystercatcher population changes. This change could be important in understanding the role of the birds in the population dynamics of the mussels.

Feeding rate varies between approximately 0·5 and 1·5 mussels eaten per bird per 5 minutes, but, as yet, no analysis on the correlates of this variation has been carried out. The birds feed at night as well as in the day, but apparently at a much lower rate because suitable mussels are difficult to detect. The early results suggest that, on most mussel beds, each bird consumes between 50 and 100 mussels per 24 hours.

Future work
In addition to the activities reported here, we are studying the winter mortality of the birds, their social behaviour, and local movements and changes in feeding behaviour and diet with age. On mussels, we are also studying factors affecting growth rates and reproductive output. These studies are planned to improve our understanding of the basic biology of the species concerned. Without this understanding, there is the danger that the simulation experiments with models will be misleading because faulty assumptions are made about the properties of the organisms involved. Study of the basic biology of both birds and mussels is, therefore, a vital adjunct to the main objective of modelling their interaction.

J.D. Goss-Custard, S. McGrorty and C. J. Reading

CRYOBIOLOGY IN CULTURE CENTRE OF ALGAE AND PROTOZOA

The ultimate aim of Culture Collections is to preserve organisms unchanged over an indefinite time. To the ordinary biologist, this aim may be considered impossible as chance is the essence of most biological processes and nothing biological is independent of time. However, at CCAP, development of methods for the long-term preservation of cultures continues. Storage under liquid nitrogen was chosen as the method of preservation. At this temperature ($-196°C$), cell survival is independent of the period of storage and biological systems are genetically stable. Thus, the selection pressures of repeated sub-culturing are avoided. It is possible to obtain high survival rates (> 75 per cent) following freezing to, and thawing from, $-196°C$, so that the probability of the selection of atypical freezing resistant mutants is low. Once cultures are stored under liquid nitrogen, there is practically no possibility of mislabelling or of bacterial contamination, and further advantages are economy of time, space and materials. The alternative method of preservation, freeze-drying, has recently been demonstrated to be mutagenic to bacteria and yeasts. The recovery rates obtained with eukaryotic cells following freeze-drying are very low and cell viability is dependent upon the period of storage. Therefore, freeze-drying is only of potential use at CCAP for the resistant resting stages such as cysts of protozoa and zygotes of algae.

Conventional methods developed for the cryopreservation of mammalian cells and bacteria give poor results when applied directly to algae and protozoa. It was therefore necessary to investigate the factors influencing the survival of algal cells following freezing and thawing. Four variables affect the response of cells to the stresses of freezing and thawing :—

1. Growth conditions before freezing.
2. Addition of cryoprotective substances.
3. Cooling and warming methods.
4. Post-thaw manipulations.

Following a period of incubation at low temperatures, some strains of algae were found to become more resistant to freezing injury. However, when the effects of cold acclimatization were investigated on a range of cells, variable results were obtained; some strains did not survive the period of cold treatment, and of those that did the freezing tolerance was either unaffected or increased. Thus, cold acclimatization, though initially promising, was not found to be of general application.

In collaboration with Dr. A. Clarke (British Antarctic Survey), studies on the biochemical mechanism of cold acclimatization were carried out. It became apparent that, for algal cells, it is not the effects of low temperature *per se* which are important, but simply the effect of reduced growth rate. Cells grown at 20°C under limiting nutrient conditions or in the presence of metabolic inhibitors become resistant to the mechanisms of injury induced by freezing and thawing, and, in all cases examined, there was a good correlation between the reduction in growth rate, increase in freezing tolerance, alterations in the degree of unsaturation of the membrane fatty acids and accumulation of lipid. However, for practical purposes, the culture of cells under reduced nutrient level, in the presence of metabolic inhibitors, or at low temperatures is not ideal as they do not reach a high cell density. Therefore, late stationary phase cultures were examined, as these are at a high cell density and are limited by several factors, especially nitrate depletion. With 20 strains of Chlorococcales, cells from the late stationary phase of culture (5–7 weeks) were found to be more tolerant of freezing than were cells from younger cultures (1 week). For all preservation studies, cells were obtained from stationary phase cultures.

Very few cell types have a high survival rate following freezing and thawing in the absence of so-called cryoprotective additives. Commonly used additives are glycerol, dimethylsulphoxide (DMSO), sugars and polymers. Unfortunately, unicellular algae are sensitive to these compounds at concentrations normally used with other cell-types. Glycerol is widely used with mammalian cells because it penetrates rapidly, but it does not freely permeate plant cells and causes plasmolysis and cell death. DMSO penetrates algal cells more rapidly, causing only a transitory plasmolysis, but causes cell death at high concentrations. For the initial preservation studies, DMSO at a final concentration of 5 per cent v/v was used, and this additive allowed the successful preservation of a large number of Chlorococcales, although the results were disappointing with other algae. Polymers such as polyvinylpyrollidone and hydroxyethyl starch were found to be relatively non-toxic to algal cells, but they were not effective cryo-

protectants. A survey of potential cryoprotective compounds was carried out using the freezing sensitive organism *Euglena gracilis* as a test system. The most effective was found to be methanol. This compound was relatively non-toxic to algal cells, and, in the case of *Euglena* and *Chlamydomonas*, is at least 10^3 times more effective as a cryoprotectant when compared on a molar basis with DMSO. Studies on the mode of protection of this additive are now in progress.

For most cell types, there is an optimum cooling rate, with damage increasing at both slower and faster rates of cooling. The actual value of this rate is dependent upon cell volume and the cellular water permeability. As these characteristics vary from cell type to cell type, it is difficult to design a freezing procedure which is applicable to a wide range of cells. Two-step cooling has been adopted as an alternative. In this method, the cells are cooled rapidly to a sub-zero holding temperature, maintained at that temperature for a period of time, and then cooled rapidly to the storage temperature. This method is simple and does not require any specialized equipment beyond a low temperature bath. On thawing, there is evidence of metabolic injury to algal cells, and the recovery is dependent upon the type of medium used. This metabolic injury is reversible, and probably indicates damage to specific enzyme systems.

While it may be over-optimistic to predict the successful preservation of all our organisms in the next few decades, it is reasonable to predict continued progress towards this practical goal, and also better understanding of the effects of freezing on organisms generally.

G.J. Morris

Research of the Institute in 1978

Introduction
This, the main section of the report, gives relatively short accounts of research projects in ITE during 1978. The main emphasis of these accounts is on projects which have been completed or in which significant progress has been made during the year. A full list of ITE projects will be found in section IV of the report.

This year, for the first time, the reports have been grouped according to the main Subdivision of ITE concerned, beginning with the three Subdivisions of Animal Ecology, namely Invertebrate Ecology, Vertebrate Ecology, and Animal Function. These reports are followed by the reports of Plant Biology, Plant Community Ecology, and Soil Science Subdivisions of the Division of Plant Ecology. Finally, the work of the Subdivisions and centres of the Scientific Services Division is described. The contribution of these Service Subdivisions is particularly important to ITE, and enables many of the research contracts and the fundamental research to be undertaken effectively and economically. Without these services, it is doubtful if any useful progress could be made.

Invertebrate Ecology

THE DISTRIBUTION AND DIVERSITY OF GROUND-LIVING ARTHROPODS IN URBAN AND INDUSTRIAL HABITATS

The study of wildlife in urban and industrial habitats, outlined in the ITE Annual Report 1974, has been continued with studies on the larger arthropods in chalk and limestone quarries and in urban gardens. The objectives have been to determine the composition of invertebrate communities in such man-made environments :

(a)   in relation to the distribution of quarries and the early stages of vegetational succession within them (Davis and Jones 1978) ;
(b)   in relation to intensity and history of urban land use (Davis 1978 and in press).

Quarries
Four sites were selected in chalk quarries in the Medway Valley of Kent, 4 in Carboniferous limestone quarries in central Derbyshire and 3 in the Ancaster area of Lincolnshire on Jurassic limestone. Sampling of the ground-living arthropods was by pitfall trapping and was divided into 9 periods extending continuously over 26–27 weeks from the first week in May to the last week in October 1975.

Altogether, 99 species of ground beetles (Carabidae), harvestmen (Opiliones), centipedes (Chilopoda), millipedes (Diplopoda) and woodlice (Oniscoidea) were recorded. They consisted largely of widespread species favoured by cultivation, species associated with dry,

sparsely-vegetated soils and species usually associated with woodlands. Several are considered to be of rather local occurrence and their presence in quarries was therefore of interest. These include the ground beetles *Stomis pumicatus*, *Amara praetermissa*, *Harpalus rubripes* and *Leistus rufescens*, the millipede *Cylindroiulus nitidus* and the woodlouse *Cylisticus convexus*. The woodlouse *Armadillidium nasatum*, whose natural habitat appears to be calcareous screes and the grassland in southern England, was found in large numbers in all 4 Kent sites and also taken in hand collections from quarries in Surrey, Dorset, Devon, Lincolnshire and Humberside. Species near the edges of their known range in England included the harvestman *Nelima gothica*, found in a Derbyshire quarry, and 3 ground beetles with mainly southern distributions taken in the Lincolnshire sites. These results suggest that quarries may facilitate the spread of certain invertebrates, as has been suggested also for several species of hawkweed *Hieracium* (Davis 1977).

An analysis of the faunistic similarity between sites showed the dominant importance of geographical location. The absence of the woodlouse *Armadillidium vulgare* from the Derbyshire quarries was particularly striking. However, faunistic differences between 2 parts of the same quarry complex in Derbyshire suggested that significant successional changes may occur in the fauna, as they do in the flora, over a period of 40 years.

Urban gardens
The effect of increasing urbanisation on insect life was reviewed with special reference to the growth of London. Quantitative changes over periods of time are difficult to obtain and so a study was made of the faunistic differences attributable to urban density and age of development. Fifteen gardens were selected in the north-west quadrant of London, ranging from the city centre to the present urban fringe between Denham and Elstree.

Pitfall trapping was again used to collect animals which have a limited mobility and were likely, therefore, to reflect local conditions. A sampling period from mid-May to mid-June was chosen in the light of experience obtained in the quarries survey as being the best single, 4-week period for obtaining maximum species diversity. Altogether, 116 species of arthropods were identified, of which 32 were ground beetles (Carabidae) and 59 were spiders (Araneae) (not identified in the quarries samples).

The degree of urbanisation was measured in relation to 3 main factors :

(a)   Distance from the city centre and nearest urban

boundary. This was considered likely to affect species through their response to climatic factors and their tolerance of pollution.

(b) The amount of open space—gardens, parks, waste-ground, etc—around each site. Measurements were made within circles of increasing size from $\frac{1}{4}$ to 2 km radius.

(c) The age and development of housing, etc.

The best linear predictors of total species richness were found to be percentage of open space (within 1 km radius) and the logarithm of the distance from the city centre. The latter implied a rapid initial increase in species as one moves away from the centre and a more gradual increase in the outer suburbs.

At the species level, there were several noteworthy records: 2 species of ground beetles, a weevil, 2 woodlice and 2 spiders had an unexpectedly widespread occurrence in London. The weevil *Barypeithes pellucidus* is associated with tree nurseries and apparently, from this study, with shrubberies. There were also several local species not previously recorded in the London area, including a millipede *Nopoiulus minutus*, which had probably been introduced with soil at some time but which has managed to survive in a small garden in Tavistock Place.

It is evident that individual gardens often have a high degree of habitat and species diversity, but they encourage the spread of synanthropic species at the expense of those with specialised requirements for more natural and undisturbed habitats.

B.N.K. Davis and P.E. Jones

References
Davis, B.N.K. (1977). The *Hieracium* flora of chalk and limestone quarries in England. *Watsonia*, **11**, 345–351.
Davis, B.N.K. (1978). Urbanisation and the diversity of insects. In: *Diversity of insect faunas*, ed. by L.A. Mound & N. Waloff, 126–138. Oxford: Blackwell Scientific for Royal Entomological Society.
Davis, B.N.K. (in press). The ground arthropods of London gardens. *Nature, Lond.*
Davis, B.N.K. & Jones, P.E. (1978). The ground arthropods of some chalk and limestone quarries in England. *J. Biogeogr.*, **5**, 159–171.

A life table study in Monks Wood, from 1972–78, using the k factor method of analysis, showed the key factor to be late larval and pupal mortality. The mortality of these stages, believed to be caused by bird predation, was dependent on the duration of the stages. At low temperatures, the stages were prolonged and mortality was high. Because of this relationship, mean June temperature was clearly correlated with adult members later in the season (r=0·898, P<0·05).

Information from entomological journals was used to identify the timing of the butterfly's spread more precisely, during the period 1930–42, which was characterised by very high early summer temperatures. Only 2 months of June in that period had mean temperatures below average for the century.

However, it is unlikely that weather conditions were entirely responsible for the spread of *L. camilla*. The abandonment of the traditional coppice management of woodland in the latter half of the 19th century, and in the early part of this century, has adversely affected a number of woodland butterflies. However, the white admiral is an exception. Where its food plant is abundant, abandoned coppice eventually becomes very suitable for this butterfly, as the trees grow up and the canopy begins to open again. The period of high June temperatures, ideal for the increase and dispersal of the species, coincided with improving habitat conditions and, together, these factors resulted in its rapid spread.

It is interesting to note that June 1972 was as cold as any in this century, yet the species survived in Monks Wood and, in 1973, was quite abundant. This suggests that, just as colonisation of the wood may have required the combination of improved habitat and favourable weather, so local extinction is likely only if habitat suitability declines, when the timing of extinction may be determined by a period of unfavourable weather.

A more detailed account of the ecology of the white admiral butterfly is given by Pollard (in press). (Plates 19, 20, Cover).

E. Pollard

Reference
Pollard, E. (In press). Population ecology and change in range of the white admiral butterfly *Ladoga camilla* L., in England. *Ecol. Entomol.*

THE EXPANSION IN RANGE OF THE WHITE ADMIRAL BUTTERFLY *Ladoga camilla* L.

The white admiral butterfly *Ladoga camilla* L., a woodland species whose larva feeds on honeysuckle *Lonicera periclymenum* L., is widespread in Europe and Asia. In England, it was largely restricted to Hampshire and adjoining counties, until it spread quite quickly in the 1930s and 40s to occupy much of the southern half of England. It arrived in the study area, Monks Wood, in the early 1940s.

ECOLOGY OF SPANISH AUCHENORHYNCHA

A primary objective of this project was a comparison of the attributes of grasslands in Britain and Spain which affected the distribution and abundance of leafhoppers (Homoptera: Auchenorhyncha). Three visits were made to the Alto Aragon region of the central Spanish Pyrenees which lies to the west of the frontier town of Jaca, Huesca province. The visits were made in June/July 1972, September 1974 and May 1977.

On each occasion, standard samples were taken from grassland vegetation at 10 contrasting sites (8 in 1972). Each sample was from 2 · 2 m² of grassland and several samples were taken from different areas of each site. In all, 98 samples were taken with a D-Vac Vacuum net. In order to place the vacuum samples within the context of the whole leafhopper fauna, the sampling programme was supplemented by an extensive programme of hand-collecting from many different types of vegetation.

Not surprisingly, the Spanish leafhopper fauna was richer than that of southern Britain. Species of the taxonomically difficult genus *Psammotettix* predominated in the samples, together with *Deltocephalus pulicarius*, which is also an abundant grassland species in Britain. The samples contained mostly species of wide distribution in Europe. However, several species of very limited distributions were also taken. Most of these appear to be restricted to the Pyrenees. The phenology of the species making up the fauna was found to be similar to that in Britain, especially on the sites at higher elevations (c. 1200–2000 m). On the lowest sites (c. 800 m), several species were found as adults in May or June and early July which in Britain occur only later in the summer. The species richness of the sites was generally high, especially so at Ordaneso, which was the outstanding site in this respect, S. Juan de la Pena, Estacion de Meterologia and Los Lecherines. There was no correlation of species richness with altitude, Ordaneso being a lowland site (c. 800 m), S. Juan at moderate elevation (c. 1200 m), and Los Lecherines the site at greatest altitude (c. 2000 m). Two sites had very low species diversity.

The grassland sites investigated were mostly very intensively grazed, and no very tall grasslands were found which could be sampled. In addition, the sample sites had much less continuous cover than British grasslands. No significant correlations were established between vegetation height, or vertical structure, and the numbers of species (S), numbers of individuals (N) or species diversity (D) in the samples. This feature contrasts very markedly with the correlations established on English chalk grasslands (Morris 1971) and limestone grasslands of the Burren, Ireland (Morris 1974). Numbers of species and of total adult leafhoppers were low in individual sites in the Pyrenees, despite the generally high species richness of its sites. The altitudinal range of the Pyrenean sites (800–2000 m) was much greater than in any group of sites studied in Britain (0–300 m). The effects of altitude on the representation and abundance of Auchenorhyncha and their phenology were especially marked.
M.G. Morris

References
Morris, M.G. (1971). The management of grassland for the conservation of invertebrate animals. In: *The scientific management of animal and plant communities for conservation*, ed. by E. Duffey & A.S. Watt, 527–552. Oxford: Blackwell Scientific.
Morris, M.G. (1974). Auchenorhyncha (Hemiptera) of the Burren, with special reference to species-associations of the grasslands. *Proc. R. Ir. Acad.* (B), **74**, 7–30.

THE ECOLOGY AND DISTRIBUTION OF PSEUDOSCORPIONS

A mapping scheme was set up in 1969 to produce distribution maps of all the British species of pseudoscorpion from lists of species occurring in each of the 10 km squares of the Ordnance Survey National Grid, and also to obtain detailed information about the biotopes in which each of the species occurs. Information on both these aspects had hitherto been unreliable. Data for a provisional atlas of distribution are being collected from both published and unpublished sources, museum collections and preserved specimens sent in for determination.

Patterns of distribution are already beginning to emerge for certain species. For example, *Neobisium muscorum* (Leach) and *Chthonius ischnocheles* (Hermann) are proving to be as common and widespread as originally thought, and *Chthonius orthodactylus* (Leach), originally thought to be restricted to the south of England, has now been found as far north as Ancaster in Lincolnshire. *Chthonius tenuis* L. Koch, previously not thought to occur north of the Thames, has been found at Wrabness in Essex and Craig y Fro in Brecon.

Some of the records which have been received for the mapping scheme have been used in a desk study of the phoretic behaviour of pseudoscorpions. Phoresy has long been known to occur in pseudoscorpions, but there is still much controversy as to whether it is predatory in nature or whether pseudoscorpions associate with other animals in order to obtain transportation. Analysis of available data suggests that phoretic behaviour in pseudoscorpions has developed as a modification of their original hunting instinct (particularly in pregnant females), which causes them to seize large animals, particularly anthropods, which come near them, at the same time allowing themselves to be carried along by them. The association is only profitable to one animal (the one being transported) and is motivated by stimuli arising only within that animal. These stimuli, released by the inadequacy of the habitat, arise from hunger and the need to find a more favourable environment for the development of the young. Probably the most important result of phoretic behaviour is the geographical distribution of the species.

More detailed research has begun into the habitat requirements and factors affecting the distribution of 2 species associated with ancient native woodland. *Dendrochernes cyrneus* (L. Koch) is the largest British species of pseudoscorpion (attaining a length of about 4 mm) and is undoubtedly also the rarest (Cover). It has only been recorded from 6 sites in England, including the 'classic' localities of Sherwood Forest, Windsor Forest and Richmond Park, and apparently lives exclusively under the close-fitting bark and in the decaying sapwood of dead or partly-dead trees (mainly oak) (Plate 22). Very little is known about its life history, except that it is probably distributed phoretically on longhorn beetles. *Chernes cimicoides*

Plate 15   Sambucus racemosa *L. Yellow ring spots in autumn foliage arising from CLRV with uncharacterised* (carla) *virus.*
Photo: I Cooper

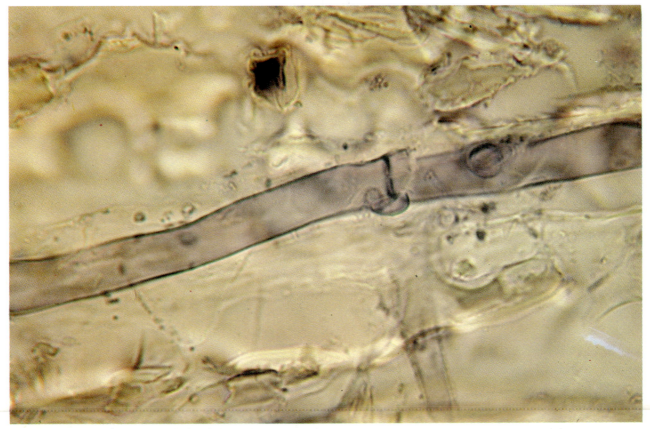

Plate 16   Hyphae of Mycena galopus (*stained blue*) *in dead plant tissue. Hyphal width: approximately* 2μm.
Photo: C Quarmby

Plate 17 Orchis militaris : *entire plant*.
Photo: L Farrell

Plate 18 Orchis militaris : *dark form*.
Photo: L Farrell

Plate 19   White admiral pupa.   Photo: E Pollard

Plate 20   White admiral 3rd instar larva and hibernaculum.   Photo: E Pollard

Plate 21 *A plantation of Sitka spruce in the Forestry Commission's Greskine Forest, Dumfriesshire. A compartment of the forest is reserved for the study of tree physiology and ecology.*
Photo: E D Ford

Plate 22 *Ancient decaying oak tree, Bilhaugh, Sherwood Forest, habitat of the pseudoscorpion, Dendrochernes cyrneus (L. Koch.).*
Photo: J Crocker, British Arachnological Society

(F.) is much more common and is quite widespread in southern England. It has not been recorded north of a line from the Mersey to the Humber. Although it has been recorded from many ancient woodland sites, where it occurs under the bark of decaying trees, it is not restricted to such habitats. Specimens have been found occasionally in leaf litter and also in the nests of ants, particularly *Formica rufa* L. and *Lasius brunneus* (Lat.).

Future work will concentrate on the collection of data on the distribution of the 25 British species and also on research into their habitat preferences (with special reference to the rarer species).

P.E. Jones

DIGESTIVE ENZYMES IN THE ANT *MYRMICA RUBRA*

This study of digestive enzymes arose from work that was being carried out at Furzebrook on caste determination in *Myrmica rubra*. Two of the paired glands in the head of this ant had attracted attention as possibly being involved in caste determination, and their functions have therefore been studied.

In the presence of a queen, large queen-potential larvae are bitten by the workers using their mandibles, and this causes earlier metamorphosis so that workers are formed rather than new queens. The mandibular glands, opening at the base of the mandibles, were found to produce a protease, but it was later shown that the mechanical process of biting was alone sufficient to induce metamorphosis.

The pharyngeal gland, which is formed as an invagination of the gut, was seen to collect the light oily fraction of the worker's food, separated off in the pharynx. This material may be passed to the larvae by regurgitation, possibly providing a vehicle for a substance involved in caste determination. However, some of the material at least was found to enter the gland cells, and, in view of this possible digestive function, the pharyngeal gland was tested for the presence of a lipase.

As little work had been undertaken on ant digestion, the study was extended to test all the head glands and the different gut sections for enzymes. Initially, simple qualitative tests for amylase, protease and lipase were employed, but thin-layer chromatography was later used for a more systematic search for a wider range of digestive enzymes.

The procedure was to dissect out the glands or gut section being investigated from a number of ant workers, incubate them in a solution of a specific substrate, and then look for the products of digestion using thin-layer chromatography. The results obtained are unlike those published for any other species of ant, but no two species so far studied have shown similar enzyme systems. Several interesting points have emerged.

No lipase was found in the pharyngeal gland, so triglycerides must be directly absorbed into the gland cells. In fact, lipase was found only in the hind-gut. Neither amylase nor maltase were found in any of the 4 glands examined, yet they occurred in the crop, the non-glandular "social stomach" of the ant. They were also found in the mid-gut, but are unlikely to have leaked forwards to the crop, which is separated by the proventriculus, acting as a one-way valve.

Most interesting, however, were the proteases. The endopeptidase found in the mandibular gland is the first indication of any digestive function for this gland. Its duct opens at the base of the mandible, so the enzyme could initiate some external digestion as the prey is masticated. Apart from this gland, there is no protease before the mid-gut. As the oesophagus and proventriculus prevent all but the smallest solid particles reaching the mid-gut, the ant must depend largely upon soluble proteins and amino-acids. However, the larvae of *Myrmica rubra* produce a copious saliva which is rich in protease, and the adult workers are known to collect this saliva. The larval saliva has also been shown to contain amylase, so perhaps it is the source of the carbohydrases found in the crop of the worker.

It is, therefore, possible that the adults make use of enzymes produced by the larvae, and the ant colony as a whole relies heavily on its brood for protein digestion. This arrangement seems to be a novel form of division of labour within the social insect colony.

A. Abbott

PHYTOPHAGOUS INSECTS DATA BANK

(This work was commissioned by the Nature Conservancy Council as part of its programme of research into nature conservation)

This data bank has been designed to collate available information about insects feeding on plants in Britain. Data are abstracted from the published literature, personal indices, and museum collections, and are then vetted, stored and processed by computer.

Analysis of the banked data provides catalogues of the food plants of the insects and lists of the insects which feed on particular plants. In addition, selected predator-prey and hyperparasitic relationships are recorded and analysed, and a controlled amount of descriptive data, such as the geographical range of the insects, is entered (if available) in the record. Figure 18 shows an example

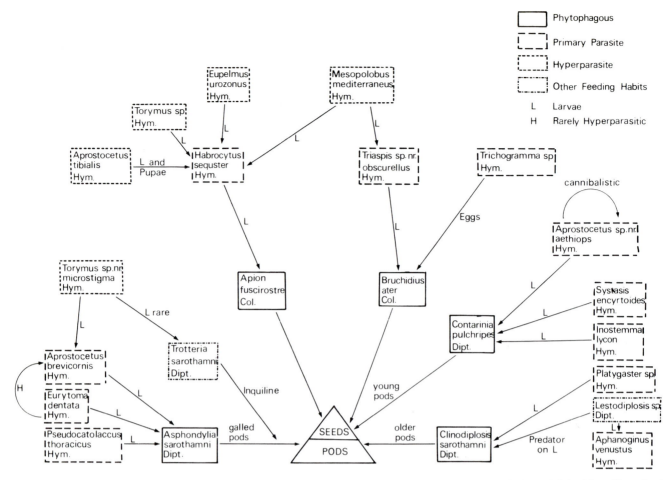

Figure 18   Insects in a complex food-chain based on the seed pods of broom (Sarothamnus scoparius). *Published from 'Grassland ecology and wildlife management'. Reproduced by kind permission of the publishers Chapman & Hall Ltd, from their book 'Grassland Ecology and Wildlife Management', E. Duffey et al. 1974. Fig. 6.5., p.14.*

of a complex food chain. (The data are taken from Waloff (1968) as presented by Duffey *et al.* (1974)).

Since the last ITE report on this project in 1976, progress has been maintained, despite considerable practical difficulties encountered in coding the data, and despite the closure of the Science Research Council's 1906A computer. The data analysed in the financial years 1975/6, 1976/7 and 1977/8 may be categorised roughly as the Hemiptera: Heteroptera (450 taxa; 1221 feeding relationships), the Coleoptera: Curculionidae (997 taxa; 2545 relationships), and a group consisting of Coleoptera: Chrysomelidae, Cerambycidae, Elateridae, Scolytidae, Platypodidae, Scarabaeoidae, Carabidae, Coccinellidae; the Hemiptera: Cicadomorpha (excluding Typhlocybinae); and Hymenoptera: Symphyta (1818 taxa; 4284 relationships). Of the total of 3265 taxa analysed so far, 3084 were 'phytophagous or partly-predacious' and 258 were 'parasitic, predacious or partly-predacious'.

Preliminary studies on this project were based on a commercially-available CDC 6600 computer running under the SCOPE operating system in London. This system was upgraded in April 1975 by the addition of a CYBER 72 computer running under KRONOS. For

reasons both of economy and dispersal of staff from London, the data bank was first implemented on the Atlas Computer Laboratory's ICL 1906A computer running under GEORGE4, and our experience of handling the data derives from that excellent system. Faced with our third change of computer system, we are hoping to gain a period of relative stability by recasting our suite of programs so that they will run both on the IBM 370/165 in the University of Cambridge and on the IBM 360/195 in the Rutherford Laboratory of the SRC. At the same time, the opportunity will be taken to improve and extend the range of analyses and facilities which can be undertaken by computer methods.

The analyses produced so far have not included the descriptive data mentioned in the first paragraph, and so the first priority will be to include these in the catalogues, and also to list explicitly the food chains found in the data bank.

The banked data, and the catalogues, are ordered taxonomically to family level, to facilitate study of the evolutionary relationships between the insects and their food plants. Experience has shown that our coding procedures can accommodate the fairly extreme variations in precision and comprehensiveness found in the available sources, but it has been more difficult than expected to maintain accuracy in the nomenclature of the species. Current checklists have been followed wherever possible, but crosschecking of synonymy and spelling in the initial sources is very demanding of manpower. The publication of any revised checklist calls for further checking, renaming or repositioning. Accuracy in coding is essential because of the unerring ease with which a computer can detect minute differences in the spelling, punctuation and layout of pieces of text or numerical data. Both manual and computer methods for maintaining nomenclatural accuracy have been devised, and it is hoped that, ultimately, checking of names and subsequent operations can be almost entirely automated. Another time-consuming manual procedure which may be automated is the extraction of the 'associated flora of the fauna of particular plant species'.

In the current financial year 1978/9, abstracting has continued while the data bank is being remounted. More data on the groups already entered have been coded, and also data on Thysanoptera, Hymenoptera: Cynipidae, Homoptera: Aleyrodidae and Homoptera: Typhlocybinae. Abstracting will continue with other groups of Homoptera, Hymenoptera, and probably also Diptera.

As information accumulates in the data bank, the range and depth of the feeding relationships will be extended, and so the value of the data will steadily increase. It is intended eventually to publish the processed data in a monograph, using computer typesetting methods to reduce proof-reading costs. In the interim period, analyses can be produced on a line-printer, and on computer output-microfiche at reduction ratios of 1/48, 1/42 or 1/24. Work on some of the general ecological and evolutionary aspects of insect/food plant relationships is also beginning as the data become more representative.

L.K. Ward and D.F. Spalding

References
Duffey, E., Morris, M.G., Sheail, J., Ward, L.K., Wells, D.A. & Wells, T.C.E. (1974). *Grassland ecology and wildlife management.* London: Chapman and Hall.
Waloff, N. (1968). Studies on the insect fauna of Scotch broom *Sarothamnus scoparius* L. Wimmer. *Adv. Ecol. Res.,* **5**, 87–208.

THE LEPIDOPTERA DISTRIBUTION MAPS SCHEME

(This work was commissioned by the Nature Conservancy Council as part of its programme of research into nature conservation)

This was the first scheme set up by the Biological Records Centre to record and map a large group (c. 1000 species) of animals. It was modelled on the techniques evolved for the BSBI mapping scheme for the British flora. The basis of the Lepidoptera Scheme is to compile, for each of the 3862 10 km squares of the Ordnance Survey National Grid (in Ireland, of the Irish National Grid), as complete a list of species of the Macrolepidoptera as is practicable. The Scheme was launched in February 1967 at the Verrall Supper (an annual gathering of some 250 of the country's leading entomologists), at which meeting the first contributors were enrolled. During 1967, the Scheme was publicised in the entomological journals, natural history periodicals, and on BBC sound radio. An exhibit was also shown at meetings of natural history societies.

As a result of this publicity, no fewer than 600 potential recorders were enrolled in 1967. Interest has continued undiminished over the years, and the number of offers from lepidopterists, both professional and amateur, to participate in the Scheme has now reached some 2600. Of these, more than half have contributed records. A network of county referees has been set up, but has proved to be only partially successful.

Feedback to the recorders has been in the form of an annual newsletter, which reports progress, and as provisional atlases. Progress reports have also been published in the entomological press. Maps have been produced to illustrate scientific papers and a complete atlas of butterflies appeared in South (1973) (figure 19). Maps of moths will appear in the *'Moths and butterflies of Great Britain and Ireland'*, a projected 11-volume work, edited by J. Heath.

Very early in the Scheme, it was realised that the literature available for identification was inadequate and that a large number of 'critical' species could not be identified satisfactorily. It was impracticable to attempt to operate a voucher specimen identification service. Lists of the difficult species were supplied to recorders and a series of 7 *'Guides to the critical species'* was published between 1969 and 1972 (figure 20). Training courses on identification techniques have been organised in conjunction with the Field Studies Council in most years, on which some 150 lepidopterists have enrolled. Three such courses were also run in the Republic of Ireland.

Recorders are instructed to visit as many different habitats as possible in each 10 km square on a sufficient number of occasions to cover the flight periods of all the species likely to occur. Records are based almost exclusively on adults, as it was not considered possible to achieve a satisfactory cover of the country in a reasonable period of time by any other method. As many species are nocturnal, recording has to be carried out both by day and by night. For the latter, a portable, battery-operated light trap, which was developed by J. Heath in the mid-1960s, has been used extensively. Particular attention has been given to nature reserves and other areas of special interest.

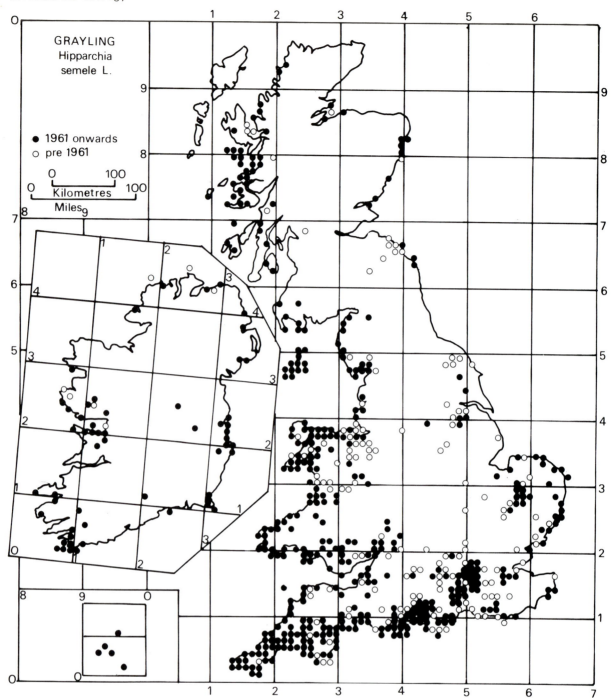

Figure 19 Distribution map of the Grayling. Reproduced by kind permission of Frederick Warne (Publishers) Ltd., from their book 'British Butterflies' by R. South, revised by T. G. Howarth, (1973).

Three types of record card are used by participants in the Scheme. To augment the field records, data from museum collections and the literature have been abstracted for the less common species. After completion, the cards are checked by the Scheme organiser and the records summarised on a 'master card', of which there is one for each 10 km square for each of 3 data classes: pre-1940, 1940–1960, 1961 onwards. Initially, the data were processed mechanically, using the Power-Samas 40-column card system, and then using a system based on the ICL Atlas computer at the Computer Aided Design Centre. Currently, the data are abstracted from the 'master cards' on to OMR forms

(OPSCAN) and processed at the SRC's Rutherford Computing Laboratory, using the method described elsewhere in this report (page 86).

Provisional maps for species in all the families except the Geometridae have now been made—a total of some 500 species. About half of these have been checked and edited by appropriate experts. The data for the Geometridae are being processed and the maps should be ready for editing in mid-1979. The data from the British Isles are also being made available to the European Invertebrate Survey (ICIS-IUBS) for in-

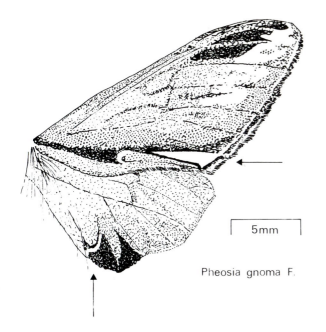

Pheosia gnoma F.

5mm

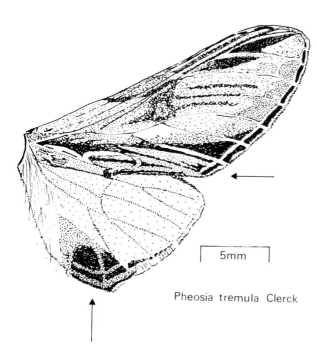

5mm

Pheosia tremula Clerck

Figure 20   Figures from Guide to the critical species, part 1, showing differences between Pheosia gnoma *F. and P.* tremula *Clerck. Reproduced Entom. Gaz. 20:93 by kind permission of the editor.*

clusion in maps for all Europe, the first 12 of which are currently in preparation.

It is hoped that a definitive atlas of the butterflies (70 species) will be published in 1980/81, and of the moths (900 species) in 1982/83, and it is expected that both will be ITE publications.

J. Heath

Reference
South, R. (1973). *British butterflies*, revised by T.G. Howarth. London: Warne.

Vertebrate Ecology

OTTER BREEDING AND DISPERSION IN MID-DEESIDE, ABERDEENSHIRE

Otter breeding and dispersion were studied for 4 years, 1975 to 1978, on 2 adjacent lochs near Dinnet, Aberdeenshire, and on a large stretch of the nearby River Dee. In October 1977, a 4-month old male otter born at, or near these lochs, was marked with harmless radioactive zinc (^{65}Zn), which appeared in its faeces (spraints) and enabled the animal's movements to be followed. It stayed exclusively at the lochs until 7–8 months old, using both lochs daily, then moved to the nearby River Dee, returning to the lochs each day, and gradually extended its range until places 58 km apart had been visited. On the river, radioactive spraints were initially concentrated together, but were later found at the extremities of the animal's range. After this otter left the lochs, it was finally recorded about 40 km upstream (having visited places 70 km apart), and it left the River Dee in June 1978.

In addition, families of unmarked otters were recorded from sightings and from measuring footprints. Prints of fore-feet less than $5 \cdot 0$ cm in length were taken to be made by baby otters, younger than 4-5 months, provided that a larger animal was also present. Between November 1974 and July 1976, at least 7 otter families were reared at the lochs, sometimes overlapping but replacing each other. This was a period of good breeding with about 10 young reared, and a mean family size of $1 \cdot 4$. From late 1976 to mid-1978, there were 5 other families, including one continuing beyond the end of the study. This was a period of poorer breeding with only 2 young reared and a mean family size of only $0 \cdot 4$. In 1974/75 and 1975/76, all winter families survived, but, in 1976/77 and 1977/78, some did not, and in the "poor" years no new babies were born until mid-summer. At the time of writing this account (November 1978), there are signs that another "good" period may be starting.

Whilst 2 or more families and other single females sometimes occurred on the 2 lochs at the same time, they may have used different parts of the lochs or the same areas at different times. In the "good" years, families stayed at the lochs on average about $7 \cdot 6$ months, compared with about 12 months in the "poor" years. No breeding holt was found. Most families were probably born in very secluded places in woods or up tiny tributaries or drains near the lochs, but at least one family each year was thought to come to the lochs from the river when the young were half-grown.

Single otters were seen on 45 and 61 occasions in the "good" years 1975 and 1976, and on 9 and 2 occasions in the "poor" years 1977 and 1978. A presumed adult male was seen in 1975 and 1976, but not in the later years. Whilst there was no information on the total number of otters present, the variation in numbers of single animals (presumably mostly young adults or pregnant females) was perhaps the biggest difference between the 2 sets of years.

In 26 km on the River Dee, about 5–6 young were reared in 3 families in November 1976–June 1977, and 1–2 in one family in January–June 1978. Four other families, 2 in winter and 2 in summer, were unsuccessful. Two families were born in autumn 1978. This was at the end of the study and their fate is unknown.

On the River Dee, spraint sites tended to be concentrated when young were known to be present, but distributed at random at other times. More spraints were sometimes found up tributaries in autumn than earlier in the year. One stretch of river had significantly fewer spraints than would be expected. This stretch was devoid of cover. The places where young otters were recorded were mostly quiet and difficult of access for people.

At the lochs, the main food was eels. Salmonids, mostly small fish, and eels preponderated in spraints from the rivers and the Dee tributaries, and in other rivers a greater variety of fish was eaten when more species were available. On average, otters took only 4·9 minutes to catch a prey item at the lochs, and fully-grown animals could probably catch sufficient for themselves to eat in about 15 minutes per day in most months. However, access to open water was restricted by ice in very cold weather, and this may have led to competition for food, failure to breed and the death of young born in autumn or early winter in seasons with much ice.

The main tentative conclusions are that the breeding rate was probably enough to maintain the otter population, but data on mortality rates are lacking. Competition for access to food may have limited breeding production, especially in and following severe winter weather. Availability of secluded places may be important for otter breeding.

D. Jenkins, R.J. Harper and G.O. Burrows

BADGER DISPERSION IN SCOTLAND

The main questions in this project are concerned with the environmental factors influencing range and group sizes of badgers in several study areas in Scotland. The project started in 1975, and a first extensive phase is due to end in 1979.

This year, we have continued to monitor range sizes, group sizes and food selection in different habitats. Range sizes are measured using radio location, whereby individual badgers are followed and observed at night,

and using colour marking of food, whereby badgers deposit markers on latrines on their territorial boundaries. Variation in range sizes was found not only when comparing different habitats, but also when comparing subsequent years in the same habitat, especially in the lower badger densities. The numbers of badgers inhabiting these ranges were established with a newly-developed technique, in which a few animals were injected with the radioactive isotope ^{65}Zn: the substance could be detected in the faeces for a few weeks, and from the proportion of radioactive faeces the number of badgers in the range could be estimated.

There appeared to be no correlation between the number of badgers in any one group and the size of the range occupied by it. In one area with relatively low badger density (approximately 3 badgers km^{-2}), there were important changes in territorial boundaries between "clans", compared with the previous year; some clans had split up, several others had joined ranges. We know little as yet about the underlying mechanisms and possible environmental correlates.

In the analysis of badger food in the different ranges (through faecal analysis), the emphasis continues to be on earthworms. Data have been collected over a two-year period (approximately 3,000 droppings), and the results are being analysed to compare study areas and badger groups with different food availability, and also to look at seasonal fluctuations.

Relations of badgers within the "clans" were studied in a captive group at Banchory, where badgers have bred successfully over the last years. Only one female out of several reared cubs, and there is evidence that overall reproductive output is suppressed through aggression between females. There are no data as yet relating this aggression to environmental variables.

H. Kruuk and T. Parish

RESEARCH ON WILD AND DOMESTIC CATS IN SCOTLAND

This study ran for 3 years from November 1975 and results are currently being analysed. The aims of the study were :—

1. To compare feeding ecology and social organisation in wildcats *Felis silvestris* and domestic cats *F. catus*. Wildcats were studied at Glen Tanar, Aberdeenshire; domestic cats were studied in 2 areas. They were either "free-ranging", when living unrestrained in the same areas as the wildcats, and in farmland in the Outer Hebrides, or "feral" on an uninhabited island.

2. To assess the influence of prey, weather and habitat on cat hunting strategies. Prey were considered in terms of their distribution (habitat

preference), availability (variation in numbers of adult, young and diseased animals) and catchability (anti-predator behaviour).

A. *Basic biology of wild and free-ranging domestic cats*
The data (diet, reproduction, parasites, identification of cats, etc.) were derived from corpses obtained from gamekeepers in and around Deeside and Glen Esk.

1. Identification of possible hybrids.
Skull measurements of known *F. silvestris* and *F. catus* were compared with those of known hybrids.
2. Reproduction
Most wildcat litters are born in May, and not in winter. Domestic cats may have several litters each year, and feral cats may breed in winter.
3. Diet
Analysis of 437 wildcat faeces deposits showed that lagomorphs (mostly rabbit) formed the bulk of the diet (91 per cent frequency of occurrence). Bone fragments and teeth in faeces, and prey remains in the field, indicated that young and myxomatosis rabbits were important prey items, taken seasonally as they became available. Other prey items included rodents (short-tailed voles and bank voles, 16 per cent), birds (mostly game species and small passerines, 15 per cent), and several infrequently-occurring items (grass, insects, shrews, anurans).

B. *Behaviour of wildcats at Glen Tanar*
This study area is a wooded valley with open moorland and high hills behind. The study involved :—
1. Describing the habitat according to the distribution of potential and preferred prey (with indices of abundance) and vegetation cover.
2. Live trapping, and assessing 'signs' (faeces, kills, tracks, scratches, lairs) of wildcats. Studying social organisation, habitat utilisation, prey exploitation, and competition with other predators (domestic cats, foxes and man) through radio-tracking.
3. Following the movements and visual/scent marking behaviour of wildcats by marking them with injections of radioactive isotopes and subsequently detecting these in their faeces, and then estimating wildcat density from the ratio of radioactive/non-radioactive faeces collected from regular transects.

Results
1. Habitat
Young forests (especially those with scrub associations) were the most important habitat, providing food (rabbits and rodents), cover for hunting, and refuge against adverse weather conditions. In summer, wildcats also lived in open heather moorlands, but they moved to forests at lower altitudes in winter in persistent snow.

2. Home ranges
These data were based on the live trapping and radio-tracking of 18 wildcats. Wildcats had relatively fixed home ranges depending on sex, age and season. Adult females occupied monthly minimum home ranges of 9–169 ha (average 75 ha) and were relatively sedentary, since all monthly fluctuations were centred on a core area which never shifted. Adult males occupied monthly minimum home ranges of 84–172 ha (average 127 ha), centred on a core area for the winter months at least. Occasionally, both adult males and females made forays of several kilometres to a new location for several days, then returning to their original area. One adult male held a fixed home range over the winter months, but when the snow melted he shifted approximately 35 km in 6 weeks before contact was lost. Juvenile females held the smallest monthly minimum home ranges (average 38 ha), which were also usually fixed on a central core area. Juvenile males were highly nomadic, and, with one exception, resided only several days in a particular valley system before moving on to the next.

3. Social organisation
Wildcats appear to be solitary, since, apart from the mating season, all radio-collared wildcats hunted and camped alone. Resident wildcats rarely visited the core areas of neighbouring resident wildcats, but the fringe areas or their home ranges often overlapped. Many faeces were deposited on conspicuous landmarks (e.g. grass tussocks and heather clumps along trails), especially near camp-sites and hunting grounds, rather than on the perimeter of the home range. In shared hunting grounds, different wildcats tended to deposit faeces in the same general areas, but individual wildcats tended to use their own particular places. Whenever 'intruders' (juvenile males) passed through a resident wildcat's home range, the rate of faeces marking appeared to increase. These data indicate that wildcats may be

territorial and actively defend parts of their home range. In support of this statement, radio-collared free-ranging domestic cats occupied different home ranges from wildcats, usually at lower altitudes.

C. *Domestic cats*
This study compared the social organisation and feeding ecology of free-ranging cats on farmland in north Uist with a population of feral cats confined to the uninhabited Monach Islands.
The underlying reasons for this study were:

1. Domestic cats exist in the wildcat study area on Glen Tanar and possibly compete and hybridise with wildcats. Early data indicated that domestic cats' social behaviour was different in many aspects; hence a study of each type in isolation seemed worthwhile.

2. On the other hand, the hunting strategies of domestic cats and wildcats appeared to be similar in many respects. Thus, the more easily observable and identifiable domestic cats provided data to help interpret the scanty data obtained on wildcat hunting behaviour.

Results
At farms, free-ranging farm cats lived communally. Their home ranges overlapped, and they shared hunting grounds with cats from the same and neighbouring communes; each cat, however, hunted alone. Solitary hunting appeared to be related to the hunting strategies employed to catch rabbits, and to differences in the dominance status of individual cats. When breeding, these cats occurred in pairs, and most litters were born and raised in the fields. Breeding pairs were more aggressive, sprayed (urine and/or anal gland secretion) more frequently, and excluded other cats from an area surrounding their dens (which were usually in the hunting grounds).

The feral cats on the island were nearly always solitary. Some appeared to be dominant and territorial throughout the year. Territories were based on rabbit stronghold areas, with subordinate cats excluded by aggressive encounters. Faeces were deposited on grass tussocks on trails within the territories, and may act as visual/scent posts reinforcing the presence of the territory-holder. In contrast, subordinate feral cats and all farm cats usually buried their faeces.

These differences in social organisation appear to be associated with the availability of winter food and access to human habitation. The basic diet of all these cats was rabbit (88 per cent of occurrence in 347 faeces deposits). Similar hunting strategies were employed by all cats, especially in catching young and myxomatosis rabbits. When catchable rabbits became less available in winter, farm cats ate more rodents or relied on food hand-outs and scavenging from the farms. Farm cats, therefore, had alternative food sources throughout the year and their ability to exploit them reflected in their flexible social organisations. On the other hand, feral cats on the Monach Islands had no adequate alternative food sources in winter. Competition for the reduced rabbit populations may have resulted in a system of exclusive territories.

L. Corbett

References
Corbett, L.K. (1978). Current research on wildcats—why have they increased? *Scott. Wildl.*, **14** (3), 17–21.
Corbett, L.K. (in press). A comparison of the social organisation and feeding ecology of domestic cats (*Felis catus*) in two contrasting environments in Scotland. In: *Proc. 1st int. Conf. Domestic Cat Pop., Gen. and Ecol.* Massachussetts, USA: Carnivore Genetics Research Centre.

RESEARCH ON SPARROWHAWKS

The research described here forms part of a wider study which aims to determine what is limiting the numbers and nesting success of sparrowhawks in different parts of Britain. The reason for studying sparrowhawks is that these birds suffer from all the adverse factors that are currently affecting other birds of prey in Britain—continuing organochlorine pesticide use, gamekeeper persecution and land use change—yet they are still sufficiently numerous in some districts to provide large samples for investigation. As various aspects of this work were described in previous reports, the account below deals only with recent findings on dispersion and social organisation.

Sparrowhawks are difficult to observe directly, so the only effective way of studying their day-to-day movements and behaviour away from the nest is by the use of radio-telemetry. This technique entails fixing a small radio-transmitter to the base of one of a bird's tail feathers. The radio is powered by a hearing-aid battery, and the whole package in a hard waterproof jacket weighs only 4–5 g, about 2 per cent of the weight of the hawk. It causes minimum discomfort to its wearer, and is shed at the next moult. Meanwhile, the radio continues to give pulsing signals for up to 3 months, enabling the bird to be found at any time, and tracked as it moves through the countryside in search of food. If the places where a bird has been found

Plate 23   Ash with typical symptoms of dieback, namely the premature loss of foliage.
Photo: C Hatton

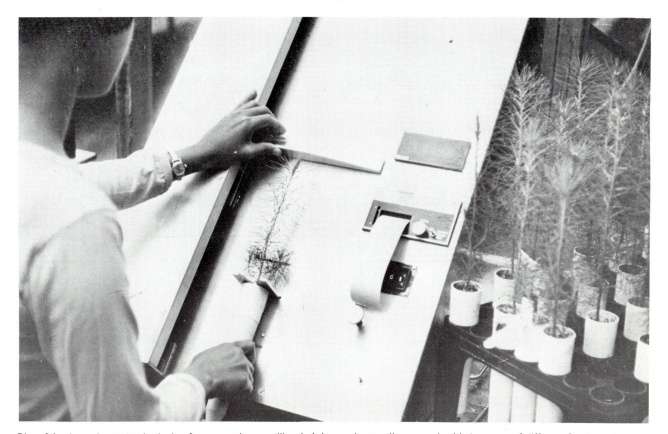

Plate 24   A semi-automatic device for measuring seedling heights and stem diameters, in this instance of different families of loblolly pine, Pinus taeda L.
Photo: M G R Cannell, while working on sabbatical leave with the Weyerhaeuser Company in Arkansas, USA.

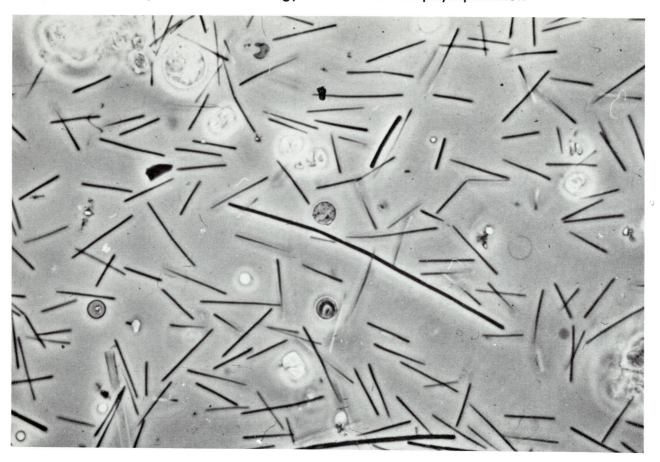

(a) *Pre-1971 plankton*

Plate 25   Rod-like cells of Synechococcus *n. sp. approximately 10μm long, and unicellular centric diatoms.*

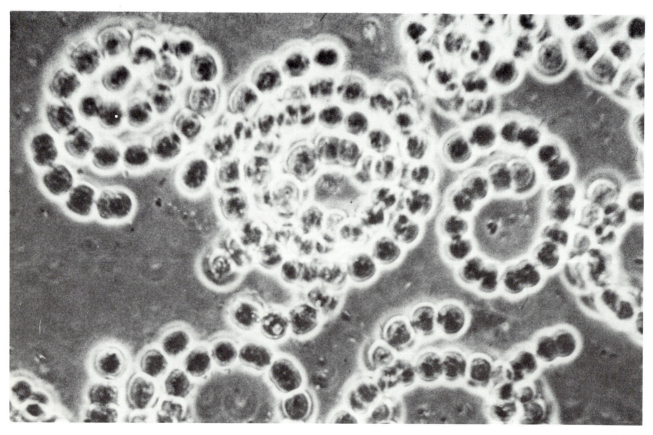

(b) *Post-1971 plankton*

Plate 26   Anabaens flos-aquae f. flos-aquae Komarek and Ettl (cell widths ca 5 μm)—September 1974.

over a certain period are plotted on a map, they can be used to provide an estimate of the bird's home range over this period, and to show which habitats the bird favours for hunting. By using different wavelengths, several birds can be studied in the same area over the same time period, enabling us to see how individuals relate to one-another in the areas they use. This information tells us about the social organisation of sparrowhawks, and helps us to define more precisely the kinds of landscapes in which they thrive best.

During the year under review, many sparrowhawks were studied in different habitats and at different seasons, and the following conclusions on their behaviour were drawn:

(a) For most of the year, sparrowhawks did not hold mutually exclusive feeding territories, but large home ranges, which overlapped widely with those of their neighbours. In general, the larger hens had more extensive ranges, and spent much more of their time in farmland and other open areas, than the cocks which were more confined to woodland. In addition, juvenile birds had much larger ranges than older birds of the same sex. This difference between adults and juveniles perhaps reflected a greater difficulty the juveniles experienced in obtaining sufficient food. The

overlap between individuals was extensive, and at least six different sparrowhawks visited certain places in the study areas during the course of a single day. For hunting, the birds preferred broadleaved woods to coniferous woods, and in farmland they made much use of tree-clumps and other patches of cover, flying from one patch to another throughout the day. Most birds had favourite roosting sites to which they returned night after night (see figure 21).

(b) Only for a brief period near the start of breeding were the home ranges of cocks mutually exclusive, as each bird hunted mainly around its nest site. However, after the young had hatched and food demands increased, the cocks began to range over larger areas again, and to overlap widely with their neighbours. The hens stayed strictly at their nests for 2 months or more, from before laying until the young were about half-grown. During this time, they depended on their mates for food. Thereafter, however, they began to range over larger areas, and often travelled greater distances for prey than the cocks. Within pairs, the hunting ranges of cock and hen bore little relationship to one another, the cock hunting more in woodland and the hen more in the open.

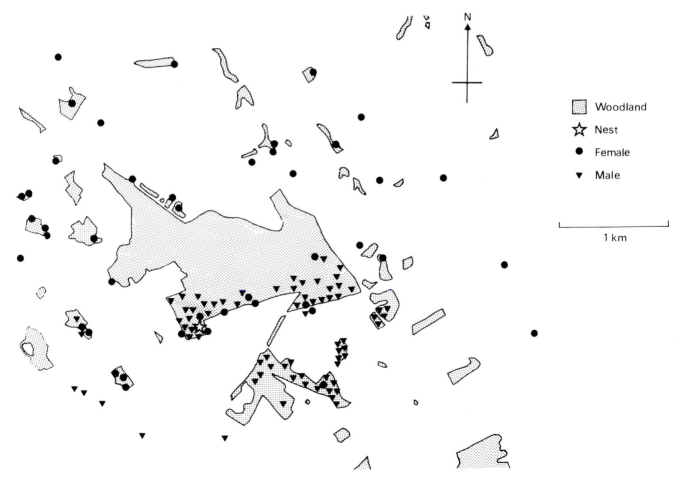

Figure 21 Hunting locations of a sparrowhawk pair in the post-fledgling period, as found by radio-telemetry. The female hunted over a larger area and in more open country than the male.

(c)   The home ranges of both cocks and hens were, at all times of year, more extensive in a large conifer forest, where prey were scarce, than in a nearby area of well-wooded farmland, where prey were plentiful. Some hens nested 1 km apart in the forest, but flew up to 9 km outside the forest to obtain prey. They commuted back and forth several times each day with food for their young, often hunting much closer to the nests of other sparrowhawks than to their own nests.

(d)   Aggressive interactions were seen chiefly in the immediate nest vicinity, and, so far as we could discover, these were the only places that sparrow-hawks defended against others of their kind.

Our overall impression was that home range sizes in sparrowhawks were influenced by (1) habitat, with larger ranges where prey were scarce than where prey were plentiful; (2) the age and hunting ability of the bird concerned, with larger ranges in inexperienced juveniles than in experienced adults; and (3) the food needs of the individual, with larger ranges when there were young to feed. As prey became scarce in winter, and sparrowhawk numbers fell, it was the juveniles that died or moved on first (Plate 8, Cover.)

I. Newton and M. Marquiss

FLUORIDE IN PREDATORY MAMMALS AND BIRDS

The aim of this study is to determine the amount of fluoride in predators, 2 feeding on vertebrates (fox and barn-owl) and 2 on invertebrates (mole and magpie), in relation to an industrial fluoride source on Anglesey (an aluminium smelter at Holyhead). Most pollutant fluoride is gaseous or particulate, and its distribution is largely determined by the prevailing winds. It finds its way into the ecosystem by dissolving in rain and soil water and so into plants. It would be expected that fluoride would then pass into animals which eat plants, both vertebrates such as voles and invertebrates such as earthworms, and from these to the predators.

Material representing all these animals was collected in 1977 at various distances and directions from the source, and also from control areas believed to be un-polluted. Analysis of various tissues was begun in 1978, and is continuing. The results so far indicate that:

1.   The concentration of fluoride in both predators and prey is inversely related to distance from the smelter, and is generally higher to the north-east than to the south-west (i.e. in the direction of the prevailing south-westerly winds). Values near the smelter reach 8,600 μg g^{-1} (mole bone). A typical non-polluted value would be about 200 μg g^{-1}.

2.   Fluoride reaches its highest concentration in bone; this was true for all vertebrate species studied. The amounts in other parts of the body vary for reasons which are not understood, and are mostly of the order of 1/100 of that in bone.

3.   Concentration of fluoride in bone varies with age. In foxes, a marked positive correlation was found between bone fluoride and age as determined by tooth sections. Adult magpies and barn-owls had about 2–3 times as much fluoride in the cranium as nestlings from the same area.

4.   Concentration of fluoride in predators is higher than in their prey. This increase was most striking in the case of moles which contained, on average, about 40 times as much fluoride in their bones as was found in the tissues of their principal prey, earthworms; but was true to some extent of all the predators.

K.C. Walton and D.C. Seel

RED GROUSE AND CAECAL THREADWORMS

(This work was supported by the Royal College of Veterinary Surgeons, and the Game Conservancy)

Grouse numbers decline about every 5–7 years and the cause has been uncertain for a long time (Wilson and Leslie 1911). During some declines, many grouse are found dead in poor condition, bearing thousands of tiny threadworms (trichostrongyles) in their intestinal caeca. Jenkins, Watson and Miller (1963) proposed that many of these birds die because they are surplus to the breeding population, and that the proximate cause of their death is irrelevant. This is an example of compensatory mortality, and their proposition is supported by the observation that some grouse in good condition contain more trichostrongyles than some of those that die. The purpose of this study has been to find whether trichostrongyles kill well-fed grouse or reduce their condition and health

The first stage in the life cycle of the threadworm begins when the eggs, which are shown in Plate 12, are passed in the caecal faeces of the grouse on to the heather moor. On warm damp days, the eggs hatch and develop within a week through two larval stages to become infective larvae (Plate 13). In damp weather, these larvae climb to the heather tips that grouse eat. Inside the grouse, the larvae grow to adults within 2 weeks and after mating with a male, which is shown in Plate 14, the adult female worms begin laying eggs at the rate of 400 per day. As they continue to pass eggs for many months and as most worms survive for over a year, the grouse seem to tolerate the worms fairly well. A moderately-infected bird will pass a quarter of a million eggs per day and so the infection could build up to very large numbers in suitable weather, if birds were at high density.

To find the number of worms in wild grouse, guts were collected from grouse which had been shot in 7 parts of the UK and the worms in them counted. Most of the worms in the grouse population are carried by a few heavily infected birds. This highly skewed distribution is common in parasite infections, and fits the negative binomial. The geometric mean number of worms per bird was about 100, but 20 per cent of young birds had none and some old birds had over 10,000. It seems that the infection increases with the bird's age, which fits the observation that the worms live a long time and are tolerated by grouse.

Although some of the captive grouse which had been dosed with infective larvae did die, most of them endured the infection. This survival also fits the observed increase of infection in older grouse. Nevertheless, infected grouse lost condition, developed a slight anaemia, and showed a small drop in serum albumen and a rise in serum globulin (Wilson and Wilson 1978).

As only territory-holders breed and less dominant grouse do not get a territory, it is possible that trichostrongyles may reduce dominance. This effect was studied by putting cocks together and finding which one of a pair was dominant in a 5-minute test and then dosing him with larvae. Dominant grouse infected with trichostrongyles maintained their ability to dominate their neighbours in spite of the physiological changes referred to above. The experiment needs to be repeated to find whether the grouse can maintain their supremacy for longer than 5 minutes.

The idea was then tested that subordinate birds were intrinsically less resistant to infection than dominant individuals, by giving them the same dose of larvae. However, the number of parasite eggs produced by subordinate birds was similar to dominant birds. So, if subordinate birds in the wild do carry more worms than territorial individuals (as postulated by Jenkins, Watson and Miller 1963), then maybe the continued stress of being dominated reduces their resistance to infection and/or they pick up more larvae because they are forced to live crowded together in grassy places, where infective larvae survive better and the birds' heather food is scarcer.

Hen grouse with trichostrongyles lay fewer eggs, and the interval between successive eggs is longer. This effect may well have contributed to the smaller clutches which hens laid during a decline in population when grouse had thousands of trichostrongyles (Jenkins, Watson and Miller 1963).

It has been noted for over a century that declines in grouse populations often occur when the birds have been numerous for a few seasons (Cobbold 1973; Farquharson 1974) and when there is a late-spring frost, or some other event reducing the food supply (MacIntyre 1918). The grouse in these experiments were fed plenty of food and yet the effects on egg-laying and physiology were like the consequences of poor nutrition. Nutrition is important to the breeding performance of grouse (e.g. Moss, Watson and Parr 1975) and possibly nutrition and trichostrongyles interact making the consequences of poor food worse. For example, the worms could interfere with food absorption and utilization and cause leakage of proteins out of the gut as happens in sheep (Symons 1969). Alternatively, malnourished grouse might be immunologically less competent to protect themselves against worms, or parasitized grouse might simply lose their appetite, which is a widespread side-effect of parasite infections in other animals.

In conclusion, trichostrongyles reduce the condition of grouse and other physiological variables usually associated with good health, although they do not kill many well-fed grouse. In the wild, these effects might contribute to declines in population, especially if associated with poor nutrition.

G.R. Wilson

References
Cobbold, I.S. (1973). Contributions to our knowledge of the grouse disease. *Veterinarian, Lond.,* **46,** 161–172.
Farquharson, R. (1974). On the grouse disease. *Lancet,* **2,** 342–343.
Jenkins, D., Watson, A. & Miller, G.R. (1963). Population studies on red grouse, *Lagopus lagopus scoticus* (Lath.) in north-east Scotland. *J. Anim. Ecol.,* **32,** 317-376.
Macintyre, D. (1918). Heather and grouse disease. *Br. Birds,* **12,** 53–60.
Moss, R., Watson, A., & Parr, R. (1975). Maternal nutrition and breeding success in red grouse (*Lagopus lagopus scoticus*). *J. Anim. Ecol.,* **44,** 233–244.
Symons, L.E.A. (1969). Pathology of gastrointestinal helminthiases. *Int. Rev. trop. Med.,* **3,** 49–100.
Wilson, A.E. & Leslie, A.S. (1911). History of grouse disease. In: *The grouse in health and in disease,* ed. by Lord Lovat, 185–206. London: Smith Elder.
Wilson, G.R. & Wilson, L.P. (1978). Haematology, weight and condition of captive red grouse (*Lagopus lagopus scoticus*) infected with caecal threadworm (*Trichostrongylus tenuis*). *Res. Vet. Sci.,* **25,** 331–336.

Animal Function

PHYSIOLOGICAL AND BEHAVIOURAL EFFECTS OF ORGANOCHLORINE POLLUTANTS

The feral pigeon *Columbia livia* var. has been used in an investigation of the hormonal and behavioural effects of organochlorine (OC) pollutants. The two major sublethal effects of the OCs observed in wild birds, egg-shell thinning and behavioural abnormalities, could both conceivably be hormonal in origin, and there have been previous reports of endocrine lesions caused by OCs (Jefferies 1975). The pigeon is a particularly convenient subject for such research because it couples striking hormone patterns with an easily quantifiable behavioural display during the period from pairing to egg-laying. Preliminary results previously reported (Dobson *et al.* 1977) on the endocrine effects of PCB (polychlorinated biphenyl) have now been extended to include behavioural observations and the effects of DDE (the metabolite of DDT) and dieldrin.

Pigeons were orally dosed with OCs and paired for a half-hour period daily. A blood sample taken at the end of the pairing period was analysed for hormones. So far, only one dose level of each OC has been tested through the pairing cycle, but a series of doses has been given over the same 14-day period to unpaired birds.

The pigeon constructs a poor nest consisting of a few twigs or straws which are normally laid on a ledge. Experimentally, the birds are presented with a bowl as a

'nest site' and 20 straws each day for construction. The bowl is emptied and the number of straws that the pigeons add in 24 hours is counted. There is considerable manipulation of the nest material by both sexes, and straws may be added to and removed from the bowl several times each day (figure 22), but in control birds there is an increase in nest building activity to a maximum at the time the first egg is laid. The nest, together with incubation of the egg, is known to stimulate the secretion of prolactin which, in turn, induces the formation of 'crop milk' regurgitated by both sexes to feed the young. PCB-treated birds showed a similar pattern of nest building to controls, but there was less activity in the later part of the cycle. The reduced interest in construction at this stage was even more marked after dieldrin treatment than after PCB, whilst the DDE-treated birds collected few or no straws and laid eggs in the empty bowl (figure 22). It is not clear whether the abnormal nest building behaviour results from a

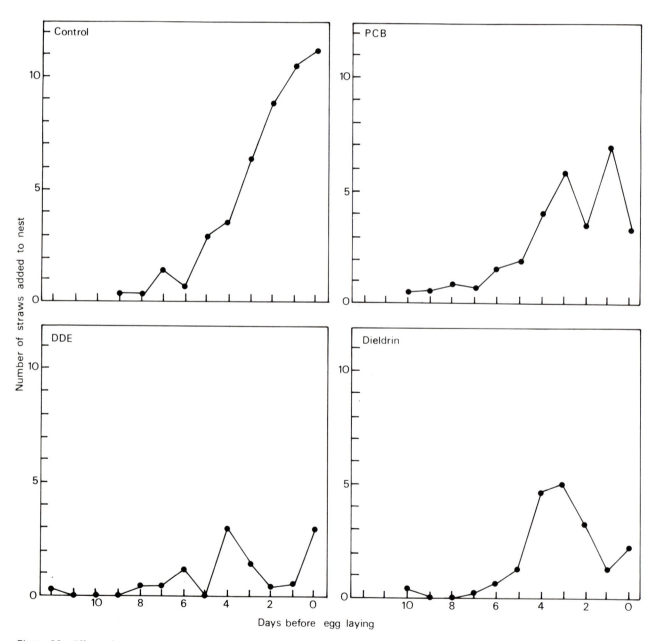

Figure 22   Effect of organochlorine pollutants on nest building activity in the feral pigeon. Each pair of birds was given 20 straws daily which could be added to a 'nest site'.

direct effect of the chemical on the brain of the bird or is an indirect effect on hormones. Nest building is assumed to be hormonally controlled to synchronise with ovulation and oviposition, but the process is not understood. It is of interest that the eggs did not show shell thinning after any of the OC treatments, though experimental shell thinning has been reported in pigeons given higher doses of DDE than were used here.

Hormonal effects were found with all 3 treatments. We have previously reported an elevation in the blood of both LH (luteinising hormone, a gonad-stimulating hormone originating in the pituitary gland) and thyroxine (the principal thyroid hormone) after PCB treatment during the egg-laying cycle (Dobson *et al*. 1977). The effect was observed only during the time when there was a normal elevation of the hormone above the baseline level of unpaired birds, and it was suggested that the brain was affected. A similar elevation in LH has now been shown after DDE and dieldrin treatment, though the effect was confined to the male. No effect on circulating thyroxine was found in either sex at this dose level of DDE or dieldrin.

Fourteen days of treatment of unpaired birds at doses up to 20 times higher than those used in the paired birds produced no significant change in circulating LH or thyroxine.

In unpaired birds, the circulating hormone levels are maintained by the operation of a feedback loop whereby trophic hormones from the pituitary gland are released under stimulation from the brain and cause the secretion of hormones from their target organ, which in turn inhibit brain stimulation of the pituitary. We can conclude from these results that high, acute doses of OCs do not affect this situation. Rising hormone levels associated with breeding require the suspension of the feedback effect of non-pituitary hormones, or the resetting of the feedback threshold. It is this neural aspect of the system which appears to be susceptible to OC interference. The differential effects in the male and female support this hypothesis. It seems unlikely that OC effects on hormone systems are significant peripherally at low doses, though chronic dosing appears to cause thyroid lesions (Jefferies 1975) and may affect other endocrine organs.

S. Dobson and N.J. Westwood

References

Dobson, S., Dobson, B.C., Murton, R.K. & Westwood, N.J. (1977). Physiological effects of organochlorine pollutants. *Ann. Rep. Inst. Terr. Ecol.*, 1976, 62–63.

Jefferies, D.J. (1975). The role of the thyroid in the production of sub-lethal effects by organochlorine insecticides and polychlorinated biphenyls. In: *Organochlorine insecticides*, ed. by F. Moriarty, 131–230. London: Academic Press.

TOXIC AND ESSENTIAL HEAVY METALS IN BIRDS

The main objectives over the past year have been:

1. To determine the inter-tissue distributions of toxic and essential metals in birds carrying high, naturally acquired, levels of mercury and cadmium;
2. To establish whether there are seasonal cycles of toxic and essential metals in bird tissues; and
3. If there are, to determine the relationship between these cycles and other seasonal cycles, such as those that occur in fat and protein. Such knowledge is required to further our understanding of the ways in which toxic metals (e.g. mercury and cadmium) might interfere with the physiological relationships between essential metals (e.g. zinc, iron, copper and manganese) and many proteins (e.g. haemoglobin, myoglobin, cytochrome oxidase, and guanylate cyclase). This area of research is central to our appreciation of the role of toxic and essential metals in many agricultural, medical, and environmental situations.

High levels of mercury and cadmium occur in several species of seabirds which breed on St. Kilda, an island group about 80 km west of the Outer Hebrides (Bull *et al.* 1977; Osborn 1978). During 1977, 10 puffins *Fratercula arctica*, 5 fulmars *Fulmarus glacialis* and 4 Manx shearwaters *Puffinus puffinus* were collected from St. Kilda in order to study the distribution of heavy metals in organs which might be adversely affected by high levels of mercury and cadmium. Zinc was also measured because it seemed likely that some effects of mercury and cadmium might be related to their ability to interfere with normal zinc metabolism.

Table 1 shows that, in all 3 species, the highest concentrations of cadmium were found in the liver and kidney, whilst substantial quantities were also present in the pancreas, gonad, and intestine. The inter-tissue distribution of zinc was similar to that of cadmium. The distribution pattern of mercury was rather more varied, with the highest concentrations being found in the liver of the 2 petrels (the fulmar and the Manx shearwater), while in the auk (the puffin) the highest levels were found in the feathers, with lower and similar concentrations occurring in liver and kidney. Cadmium was virtually absent from brain, blood and feathers, although these 3 tissues contained appreciable quantities of zinc, and sometimes of mercury.

All of the birds were apparently healthy when caught, and the puffins and fulmars were breeding (the breeding status of the Manx shearwater could not be determined). This suggests that these birds may be unaffected by the high levels of mercury and cadmium in their tissues, even though some of the levels were higher than any previously reported for wild vertebrates. One fulmar contained 480 mg kg^{-1} (dry weight) cadmium in its kidney. This apparent paradox can be partly resolved by the knowledge that much of the cadmium

Table 1.   Metal levels in pelagic seabird tissues

Figures are mean values (mg kg⁻¹, dry weight). For certain organs some of the samples contained too little metal for it to be detected. In such cases, the mean given in the table is that calculated from the samples which contained sufficient metal for analysis to be possible. The true mean is thus less than the calculated value and is shown as such.
ND=no metal detected in any of the samples of the organ.
Limits of detection: Hg, <1.25 μg per sample; Cd, <2.5 μg.

Tissue	Puffin			Manx shearwater			Fulmar		
	Zn	Cd	Hg	Zn	Cd	Hg	Zn	Cd	Hg
Liver	118	20	5	141	16	10	364	49	29
Kidney	164	114	5	176	95	5	310	228	13
Pancreas	191	22	<2	348	16	2	643	52	2
Gonad	178	17	2	157	16	2	113	9	4
Fore-gut	167	9	1	187	6	2	246	22	<2
Hind-gut	149	13	<1	167	<10	<3	133	<20	<4
Pectoral muscle	47	<2	<2	46	3	1	58	3	2
Heart muscle	100	<2	1	117	2	2	90	<2	2
Gastronemius muscle	133	<3	<2	–	–	–	–	–	–
Blood	76	ND	<2	102	ND	<5	85	ND	<2
Brain	56	ND	<2	71	ND	<6	57	ND	<3
Skin	38	ND	<1	13	2	<1	16	<2	1
Feather	108	ND	8	87	<1	1	97	<1	3

from Osborn, Harris and Nicholson (in press.)

in fulmar liver and kidney is bound to a metallothionein-type protein which may detoxify the cadmium (Osborn 1978). However, cadmium is present in appreciable quantities in both gonad and pancreas (2 organs whose function can be disrupted by cadmium) and since it is not certain that cadmium is bound to metallothionein in these tissues, it is possible that this high, naturally-acquired, concentration of cadmium is having some adverse effect on the animal's performance.

Assessing the toxic significance of metal levels in seabirds is made especially difficult because samples of many species can only be obtained during the breeding season, since the birds spend the rest of the year widely dispersed across the open ocean. For the same reason, it is difficult to determine whether the concentrations of toxic metals are always at the levels reported here, or whether they are ever higher or lower (a point which is of concern in certain types of environmental monitoring programmes).

Such subjects as these can be investigated only by working with a common terrestrial species, and, accordingly, essential and toxic metals were measured in livers of starling *Sturnus vulgaris*, together with the levels of fat and protein present in this organ. Finnish

workers (Haarakangas *et al.* 1974) had already obtained evidence from sparrows *Passer domesticus* which suggested that there were seasonal cycles in the levels of some of the essential metals (when analytical results were expressed in terms of mg metal per kg dried tissue). Not only did the starling study confirm this, but it also demonstrated that there were seasonal cycles in the concentrations of toxic metals as well. However, the starling study also showed that there were marked variations in the amount of fat and protein in the liver, and that a different pattern of seasonal cycles was obtained if these changes were accounted for. Indeed, if results are expressed in terms of mg metal per kg protein in the tissue (protein being the part of the tissue to which the metals bind), then many of the essential metal cycles seem to disappear leaving only the toxic metal cycles—although this may be an over-simplified view (Osborn in press). Figures 23 and 24 show the seasonal cycles in mercury and cadmium. Generally, the same pattern is obtained for these 2 metals no matter how the results are expressed. Peak cadmium levels are attained during moult, when mercury levels are lowest. The highest mercury levels are found in autumn when mercury-dressed grain forms a substantial part of the starling's diet.

The starling results indicate that, since such marked

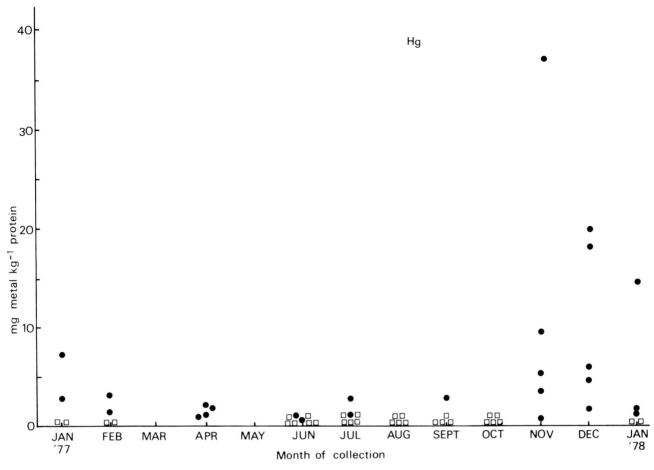

Figure 23   Seasonal cycle in starling liver mercury concentration. □ = *no mercury detected* (<1.25 µg per sample).

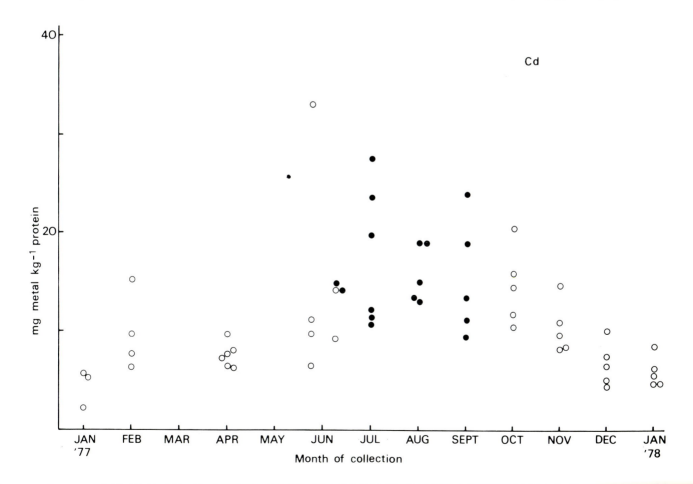

Figure 24   Seasonal cycle in starling liver cadmium concentration. ● = *birds in moult;* ○ = *birds not in moult.*

seasonal cycles exist, great care is needed if one is attempting to use animals as part of a programme to monitor environmental pollution, or if one wishes to assess the toxic significance of residues found in animals collected at only one time of year. Furthermore, they show that moult is a time of year when birds may be particularly prone to accumulating cadmium. If birds do accumulate cadmium because they are near to, or in, moult, and not because they are exposed to naturally or industrially contaminated food, then the starling results could help explain the apparent differences in the metal levels of the 3 seabird species, since the puffin, fulmar and Manx shearwater moult at different times of the year.

D. Osborn

References
Bull, K.R., Murton, R.K., Osborn, D., Ward, P. & Cheng, L. (1977). High levels of cadmium in North Atlantic seabirds and seaskaters. *Nature, Lond.*, **269**, 507–9.
Haarakangas, H., Hyvarinen, H. & Osanen, M. (1974). Seasonal variations and the effect of nesting and moulting on liver mineral content in the house sparrow *Passer domesticus. Comp. Biochem. Physiol.*, **47A**, 153–163.
Osborn, D. (1978). A naturally occurring zinc- and cadmium-binding protein from the liver and kidney of *Fulmarus glacialis*, a pelagic North Atlantic seabird. *Biochem. Pharmac.*, **27**, 822–824.
Osborn, D. (in press). Seasonal changes in the fat, protein and metal content of the liver of the starling *Sturnus vulgaris. Environ. Pollut.*
Osborn, D., Harris, M.P. & Nicholson, J.K. (in prep.). Comparative tissue distribution of mercury, cadmium and zinc in three species of pelagic seabird.

HEAVY METALS IN WADERS

Potentially-toxic heavy metals, notably lead, mercury, and cadmium, have accumulated in the estuarine sediments of industrialized river systems. The metals find their way into invertebrate animals, some of which have evolved a high tolerance, and these animals are then eaten by birds. Large numbers of sandpipers and plovers spend the winter in polluted estuaries around the British coast, and there has been some concern that the heavy metals will affect them adversely.

In conjunction with a Durham University research group, headed by Dr. P.R. Evans, a study has been made of dunlin *Calidris alpina* (*alpina* race) and knot *Calidris canutus* in the Tees Estuary in north-east England, where they remain through the winter. Paradoxically, 'Teesmouth' is highly contaminated with metals and other pollutants, yet it supports a large (and apparently flourishing) wader population.

At intervals during 1977 and 1978, small samples of dunlin and knot were taken from the large catches being collected by canon-netting at communal roosts for other ecological studies. The zinc, mercury and cadmium concentrations of each specimen's liver, kidney and brain were determined by chemists at Monks Wood. These analyses showed that there was marked seasonal variation in the concentrations of the 3 metals in liver and kidney tissue. The changes can hardly be explained by passive accumulation, since there was little change (if anything, a slight fall) in metal levels during the early winter months (figure 25). There was then a rapid increase in, or around, February, followed by a decline before the birds' departure for the arctic breeding grounds in May. Throughout their stay in Teesmouth, dunlin and knot feed on small molluscs and worms, and no seasonal changes in diet were noted that could explain the seasonal changes in the birds' metal loads. The changes can be explained, however, by assuming a natural and regulated cycle of zinc levels.

Zinc is known to be an essential micro-nutrient for domestic birds, and presumably so for wild birds. It is involved particularly in the formation of healthy skin and skin derivatives, and an adequate supply of the metal must be available for proper moult. Zinc-deficient poultry undergo abnormal moults that would prove fatal in any wild bird.

Finnish workers (Haarakangas *et al.* 1974, cf. previous article) have found a distinctive zinc cycle in wild house sparrows *Passer domesticus*, with zinc accumulation prior to moult; the females also accumulated zinc prior to laying, presumably to furnish their eggs with the metal. A seasonal zinc cycle has also been found in the starling *Sturnus vulgaris* (Osborn in press, cf. previous article), again with maximum zinc levels prior to moult.

In the dunlin and knot from Teesmouth, zinc accumulation occurred in late winter, just before the rapid pre-nuptial moult into breeding plumage. Dunlin sampled in May (after this moult, but before leaving for the breeding areas) had lost much of the zinc stored in the liver and kidney. What happens while these waders are in the arctic is not known, but, by analogy with the sparrow and starling results, a further zinc peak would be expected after breeding, when the birds begin their post-nuptial moult around July. In the females, a third period of zinc accumulation would be expected in June, prior to egg-laying. Supposedly, a regulatory mechanism exists whereby an individual meets its essential zinc requirements. When zinc accumulation is required a specific 'zinc appetite' could lead the bird to concentrate on zinc-rich items of food, or there could be an adjustment in zinc uptake from the gut and/or zinc excretion.

As figure shows, seasonal changes in mercury and cadmium paralleled the zinc changes, though the amounts of these 2 metals were always far smaller than those of zinc. It is suggested that the cyclical changes in mercury and cadmium are an inadvertent consequence of the natural zinc cycle. They accumulate when an individual's physiology is geared to metal (i.e. zinc) intake and retention, and decline when it switches to metal reduction.

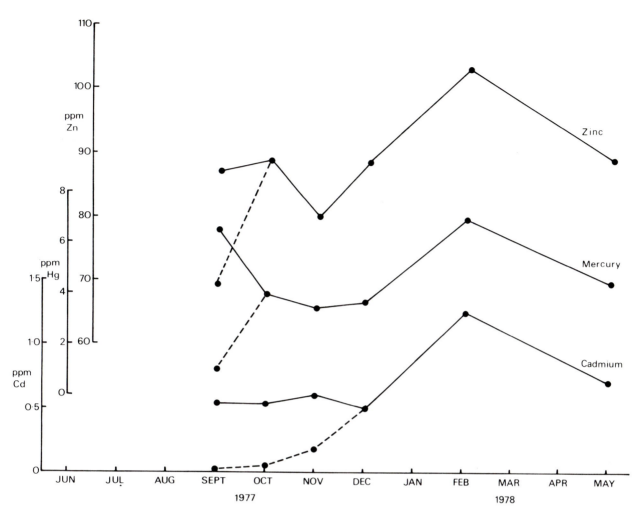

Figure 25 Changes in the mean metal concentrations in the liver of alpina *race dunlins collected in Teesmouth.
Note: Separate means given for first-winter (------) and older (———) individuals only in months when significantly different.*

A number of important questions remain: "Does the accompaniment of mercury and cadmium adversely affect the essential zinc cycle?" and "While in the body, do mercury and cadmium have toxic effects?" On *a priori* grounds, it would seem that these metals are not harmful at present. First, the numbers of waders are high in some of the most polluted British estuaries, and they are not obviously declining. Second, although industrial pollution is a recent phenomenon, like all maritime creatures shore-birds have been exposed to these metals throughout their evolution, especially in regions of high natural mineralization.

On the basis of this work, it is concluded that there is no reason to believe that the high levels (by normal toxicological standards) of mercury and cadmium found in some waders are harmful to them; and since, hopefully, the pollution situation in estuaries will gradually be improved for amenity and economic reasons, a heavy metal situation with which waders cannot cope should not arise.

P. Ward

HEAVY METALS IN THE RIVER ECCLESBOURNE, DERBYSHIRE

The Ecclesbourne has a high natural contamination with lead, cadmium and zinc. Lead and cadmium are of current interest as potential pollutants from man's activities which have no known essential biological function. Zinc is an essential element, but it can also be a pollutant in fresh waters. The Ecclesbourne appears to be biologically quite a healthy river, and a detailed study is being made of it, in conjunction with laboratory experiments that will be made using a continuous-flow system.

The underlying rationale for this study is that it should help to develop our ability to understand and predict the effects of pollutants in ecosystems. Flowing fresh waters are particularly attractive for such studies: they are relatively simple ecosystems in terms of number of species, direction of movement, and sharpness of their boundaries. Moreover, from a practical point of view, many of the most acute pollution problems occur in fresh water.

Clearly, the first need is to study the distribution of these three metals in the abiotic environment. They are derived from mineral veins that cross the valley near the head of the river. The very first step is to study the distribution of metals in the sediments along the length of the river. A somewhat similar pattern exists for all 3 metals, with peak concentrations near the source, and a relatively low constant concentration along the length of the flood-plain downstream. It may be that this pattern simply reflects the transport of particles down the river, but the effects of particle size and of organic content are also being investigated. To obtain more precise information on the distribution of metals along the river requires a more sensitive sampling technique. Such a technique is being developed—the essential feature appears to be that the bulk of the metals occurs in a small proportion of the sediment particles.

Studies have started on the amounts of metal in solution and suspension. The first problem to be encountered was that water samples stored for some time before analysis can give unreliable results for some ions. ITE's Subdivision of Chemistry and Instrumentation is now studying this problem. As soon as time permits, more detailed studies will be made of the fluxes to and from the water and suspended solids, which will be linked closely to experiments in the continuous-flow system.

These questions of amounts and kinetics of metals in the abiotic environment are of interest in their own right, but their biological significance is that they affect the amounts found within organisms. Preliminary analyses are under way for several species before a proper study is designed to investigate the effects of site, species and time of year on the amounts found within organisms. Results so far indicate that a compartmental type of approach will be useful, and that it should be possible to integrate this study with results from the continuous-flow system.

F. Moriarty, K.R. Bull and H.M. Hanson

TOWARDS A BETTER UNDERSTANDING OF THE WILD RABBIT

There is little doubt that the numbers of wild rabbits in Britain are steadily, albeit slowly, increasing, despite the continuing presence of myxomatosis. Attenuation of the virus's virulence and some enhanced genetically-determined resistance are believed to be responsible for the increase. Concerned by this growing rabbit problem, ITE has instigated research designed to increase our knowledge of the ways in which this pest species interacts with its environment. It is not envisaged that topics of purely economic importance, such as the assessment of rabbit damage, will be pursued, but rather that research of a fundamental nature will be

undertaken to provide an adequate theoretical foundation on which to base future applied work. It is proposed to study those physiological and behavioural processes which determine the way individuals and populations live in the wild and how seasonal and diurnal cycles of body function and nutrition can affect performance. These physiological cycles, affecting such processes as fat storage, moult, breeding season, etc., are ecologically adaptive and are controlled by hormonal cycles. Studies of the underlying mechanisms which control these physiological rhythms are possible now that sensitive and specific radio-immunoassay techniques are available for identifying and monitoring circulating hormones in the blood.

The incidence of rabbit infestation depends primarily on the influence of environmental factors, both physical and biotic, on the reproductive potential of the individual. The reproductive potential of the wild rabbit is extremely high and, when 2 breeding pairs were allowed to breed in isolation, 9 and 10 litters of 5 young were produced in 12 and 12 months, respectively. Normally, social contact and intra-specific competition, however, greatly reduce this potential productivity. In order to understand the influences of each environmental variable on reproductive output, captive wild rabbits are being studied in an enclosure erected on a 0.81 ha site at Monks Wood (Plates 00). Within the enclosure, animals are kept (i) in isolation from social contact, (ii) in single family units, and (iii) in colonies of different densities: a methodical investigation of the effects of individual variables can then be undertaken. Only in the final analysis will the relative importance of each environmental factor be realised, but this approach should provide the information necessary for understanding how breeding success of a rabbit population is regulated. Monitoring of a free-living rabbit population and the determination of its population dynamics will allow direct comparison with the enclosure studies (Plates 10, 11).

D.T. Davies

Plant Biology

WITHIN-CLONE VARIATION IN TROPICAL TREES

Considerable experience has been gained during the past 5 years in growing and propagating tropical trees in specially-adapted glasshouses in Scotland. Many tropical species can be readily propagated vegetatively, a development of importance for forestry practice and research. Inherent effects on growth can now be readily separated from those primarily attributable to environmental factors. In the joint UK/Nigeria projects on *Triplochiton scleroxylon*, sponsored by the Overseas Development Ministry, and coordinated by Professor F.T. Last, 0·2 million rooted cuttings have now been planted in field trials in Nigeria, with appreciable differences developing after 2–3 years. The nature of these differences needs to be understood if methods of vegetative propagation are to be exploited rationally. At least 2 types of variation can be recognised among clones, one attributable to the physiological condition of cuttings and the other to their positions of origin on stockplants.

There appear to be short-term 'carry-over' effects comparable to the "c" effects in seedling populations. Buds and roots may grow more slowly on some cuttings than on others, perhaps because of differences in apical organization, amounts of endogenous hormones or storage reserves. After rooting, however, growth becomes progressively more influenced by current conditions. In experiments planted at Gambari Forest Reserve, near Ibadan in Nigeria, there were, for the first 4–8 months, significant effects on plant height attributable to 'carry-over', but thereafter these effects disappeared, although clonal differences persisted.

Whereas cuttings originating from the upper branches of stockplants (decapitated and grown obliquely to stimulate branching) produced plants with persistently non-vertical shoots, others, from basal shoots, developed normal vertically-growing plants (plate 9). The former cuttings are, for obvious reasons, undesirable

and the management of stockplants must increase the supply of cuttings of the 'vertical' type.

Cuttings from woody adult plants are often more difficult to root than those from seedlings, an effect attributed to the onset of maturity. Do the differing growth forms shown in plate 1 reflect changes associated with this phenomenon, or has phase-change been accelerated by some aspect of the environment? On the other hand, the non-vertical habit may be related to the persistence of 'branch' rather than 'mainstem' characteristics. In Scotland, a greater proportion of cuttings from vertical plants root, and root sooner, than those of non-vertical plants (Longman 1976). It has already been shown that the rootability of cuttings is strongly influenced by propagating-bed temperatures, amounts of auxins (Leakey et al. 1975), the orientation of stockplant main stems relative to gravity (Bowen et al. 1977), and the mineral nutrition of stockplants (Leakey and Longman 1977). The rooting ability of lateral shoots is also altered by the way in which stockplants are pruned (figure 26). When 20 nodes were removed from 30-node plants, over 80 per cent of the cuttings developing from the remaining nodes rooted in 8 weeks; when only the shoot apex and 1 expanding leaf were removed, 20 per cent of the cuttings rooted. In another experiment,

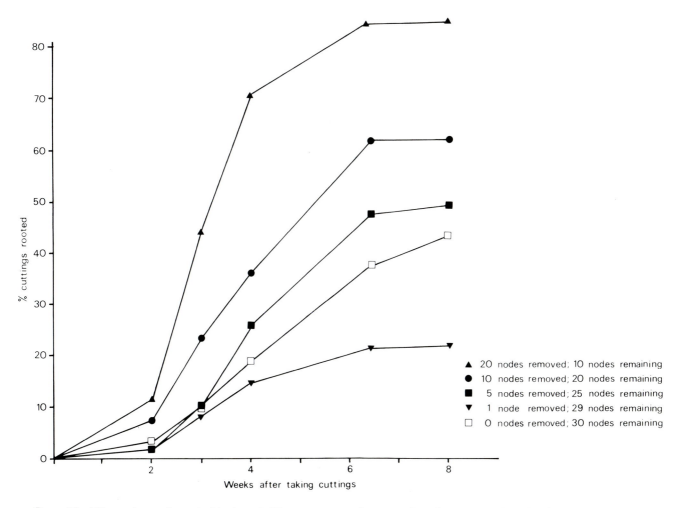

Figure 26 *Effects of removing apical buds and different amounts of stem on the subsequent rooting of leafy single-node cuttings of* T. scleroxylon. *All stockplants were left with five leaves following decapitation.*

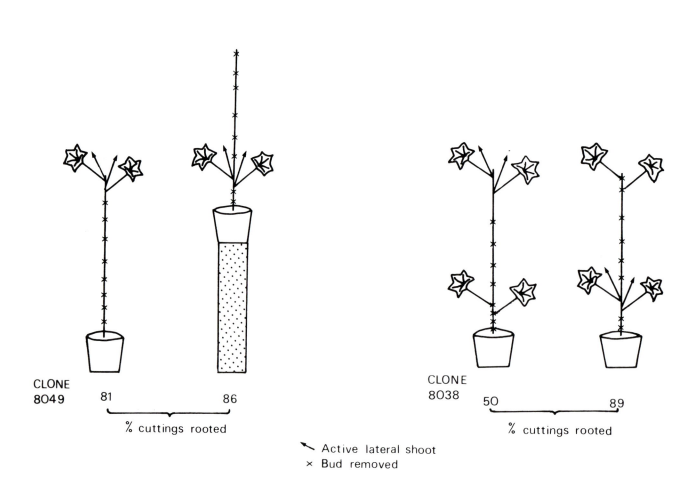

Figure 27 *Effects on the subsequent rooting of leafy single-node cuttings of* T. scleroxylon *of keeping basal shoots in the same environments as apical shoots.*

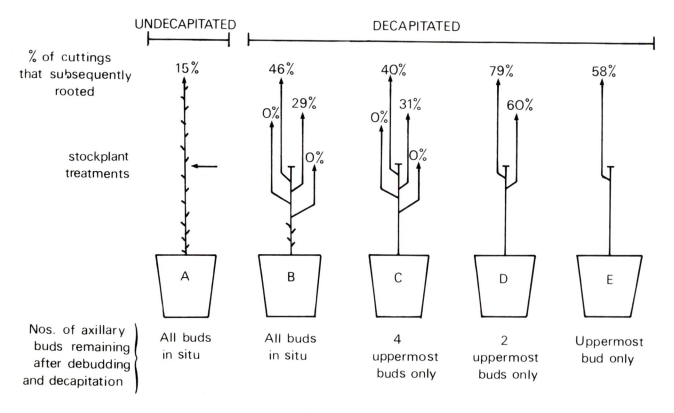

Figure 28 *Effects of controlling numbers of lateral shoots per stockplant on the subsequent rooting (%) of leafy single-node cuttings of T.* scleroxylon. *All stockplants were left with four leaves following decapitation.*

the uppermost nodes were removed ; 2 axillary buds and 2 leaves were left in place, either near the top or near the bottom of the main stem. Cuttings from both treatments rooted equally successfully (figure 27a), with little or no evidence of the usual decreasing rooting ability with increasing plant height (figure 27b).

Severely-pruned plants understandably produce fewer lateral branches than more lightly-trimmed stockplants. Earlier, it was found that the rooting ability of cuttings was associated with numbers of competing shoots (Leakey and Longman 1977). In the third experiment of the series, it was confirmed that stockplant pruning enhanced the rooting of cuttings (figure 28). The production of uniform cuttings was achieved in practice by restricting lateral growth to 2 buds per stockplant ; further restriction adversely affects numbers of cuttings.

R.R.B. Leakey and K.A. Longman

References
Bowen, M.R., Howland, P., Last, F.T., Leakey, R.R.B. & Longman, K.A. (1977). *Triplochiton scleroxylon* : its conservation and future improvement. Forest genetic resources information (FAO) No. **6**, 38—47.
Leakey, R.R.B., Chapman, V.R. & Longman, K.A. (1975). Studies on root initiation and bud outgrowth in nine clones of *Triplochiton scleroxylon* K. Schum. *Proc. symp. variation breeding systems of Triphlochiton scleroxylon K. Schum.*, 86—92. Ibadan : Forestry Research Institute of Nigeria.
Leakey, R.R.B. & Longman, K.A. (1977). Root and bud formation in West African trees. *3rd Ann. Rep. UK Overseas Development Ministry, Research Scheme R. 2906.*
Longman, K.A. (1976). Some experimental approaches to the problem of phase-change in forest trees. *Acta Hort.*, **56**, 81—90.

PHOTOPERIODIC CONTROL OF PINE FLOWERING

With the vegetative propagation of selected clones having a propensity to flower allied to the availability of controlled environment cabinets, it is now possible to hasten our understanding of the flowering of forest trees. When 2-year old plants of *Pinus contorta* were subjected to 4 different environments for 11 weeks in the factorial combinations of (i) cool (15°C/8°C night) or warm (23°C/15°C night) temperatures and (ii) short (10 hours) or long (19 hours) days, it was found that short days encouraged the development of 6 times as many females cones as occurred in long days (figure 29) This evidence of a photoperiodic effect was independent of temperatures and applied to more than one clone. The effect was attributable to an increased proportion of trees producing female cones and to a greater number of cones per tree. Male cones were more numerous in cool conditions than in warm, none being formed in the long day, warm treatment (figure 30). This observation is contrary to the generally-held belief that warm temperatures promote flowering in forest trees, but may possibly be an indirect effect of competition among different meristems for photosynthates whose produc-

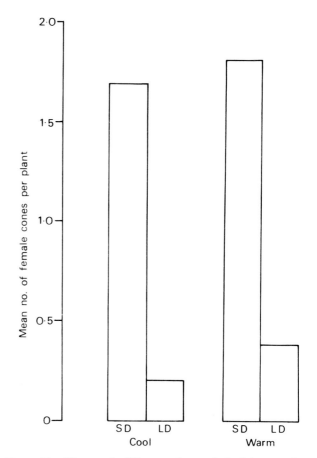

Figure 29 Effects of different photoperiods (short v. long days) and temperature regimes (warm and cold) on the production of female cones of Pinus contorta.

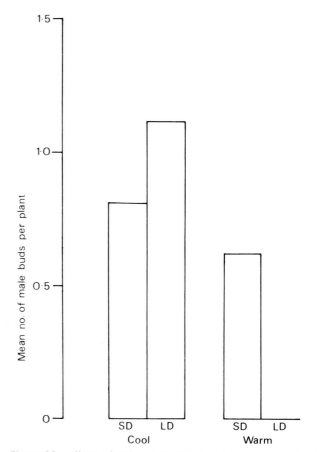

Figure 30 Effects of different photoperiods (short v. long days) and temperature regimes (warm and cold) on the production by Pinus contorta *of buds with some male cones.*

tion was limited by the moderate light intensities experienced in the cabinets. In 1978, when grown in relatively cool glasshouses, a majority of the plants again produced male cones.

Although the Pinaceae seem relatively unresponsive to GA$_3$, which stimulates flowering in the Cupressaceae and Taxodiaceae, flowering in the P. contorta has been reported to be promoted by a GA$_{4/7}$ mixture (Pharis et al. 1975). However, this response was not confirmed when clonal material was injected with GA$_{4/7}$, although the hormone mixture doubled the height of the 1977 buds and significantly lengthened shoot growth in 1978, some of them with 2 sets of lateral buds.

Different day-lengths affected vegetative growth differently. After 4 weeks of short days, the current year's needles ceased elongating, whereas they were still lengthening after 11 weeks of long days, a result agreeing with that reported for P. sylvestris by Wareing (1950). However, in P. contorta, the succeeding year's growth was also affected, with short days increasing the incidence of shoots with 2 sets of lateral buds. As a general rule, the warmer temperatures did not increase vegetative growth, but there were suggestions of possibly significant interactions. Thus, shoot growth was 40 per cent greater in 1978 in plants which had been subjected to long days and cool temperatures in 1977.

These results illustrate how significant effects of individual environmental factors, genotype, and plant hormones and their interactions can be studied in controlled environments with relatively uniform clonal material. The effect of short days in increasing female cone production in P. contorta may explain persistent reports that pines grown substantially further south than their latitude of origin tend to produce female cones more freely than plants which have been moved farther north. Male cone production tends to be affected in the opposite way, but, here, it may be expected that the cooler temperatures experienced when plants are moved north could possibly be responsible for the increase in male cones. The knowledge gained from these studies should enable practical flower-inducing techniques to be developed so that tree-breeding can be hastened and seed supplies obtained from the most promising parent trees (Longman 1978), which may well include many of those which normally exhibit little or no flowering.

K.A. Longman

References
Longman, K.A. (1978). Control of flowering for forest tree improvement and seed production. *Scient. Hort.*, **30**, 1–10.
Pharis, R.P. & Kuo, C.G. (1977). Physiology of gibberellins in conifers. *Can. J. For. Res.*, **7**, 299–325.
Pharis, R.P., Wample, R.L. & Kamienska, A. (1975). Growth, development, and sexual differentiation in *Pinus*, with emphasis on the role of the plant hormone, gibberellin. In: *Management of lodgepole pine ecosystems*, ed. by D.M. Baumgartner, 106–134. Pullman: Washington State Univ. Press.
Wareing, P.F. (1950). Growth studies in woody species. II. Effect of day length on shoot growth in *Pinus sylvestris* after the first year. *Physiologia Pl.*, **3**, 300–314.

EARLY GENETIC EVALUATION OF TREES

Tree breeders would like to recognize fast-growing genotypes of trees at the seedling stage in order to shorten the generation time, or at least decrease numbers of genotypes that need to be evaluated in long-term forest trials. In advanced tree breeding programmes, such as that established with loblolly pine *Pinus taeda* L. in south-eastern USA, there are many single-tree families which have already been evaluated in forest trials for many years. These families can be grown again from seed and used to test ideas on seedling evaluation.

During a year's secondment to the Weyerhaeuser Company, seedlings of 16 families of loblolly pine were grown in specially-designed containers, their growth characteristics subsequently being correlated with mean stem volumes of the same families in forest trials in North Carolina. Seedling characteristics included seed size effects, water stress responses, root-shoot allometry and effects of temperature and photoperiod on dates of bud-set.

In collaboration with M.R. Ingram, a device was built to enable numerous height measurements to be made daily (Plate 24). On this device, the up and down movements of the arm are translated into electric signals which are transmitted to a visual display panel and printer.

Rates of seedling height growth were influenced by family differences in seed size until seedlings were about 140 mm tall. Thereafter, and until bud-set, growth rates of well-watered seedlings became positively correlated with their long-term field performance, especially for families tested in the field on poorly-drained sites. For families tested on better-drained sites, correlations with seedling height growth rates were significant only when seedlings were grown under conditions of mild water stress: families which grew fastest in the field also grew fastest as seedlings when under mild water stress. The fast-growing families also tended to have large root-to-shoot relative growth rates, as shown for seedlings grown in 2 contrasting media by regressions of shoot on root dry weights. This suggests that fast-growing genotypes on better-drained sites avoided water stress by producing extensive root systems. About 90 per cent of the variation in field performance, after seed size effects had disappeared, could be accounted for by multiple correlations with seedling height growth rates (in given water stress environments) and shoot-root dry weight regression coefficients (Cannell et al. in press).

If these findings are substantiated, there is a possibility of developing a simple, cheap method of screening loblolly pine families for growth rates on individual trees at the seedling stage.

M.G.R. Cannell

Reference
Cannell, M.G.R., Bridgwater, F.E. & Greenwood, M.S. (in press). Seedling growth rates, water stress responses and root-shoot relationships related to eight-year volumes among families of *Pinus taeda* L. *Silvae Genet.*

VIRUSES OF TREES

Cherry Leaf Roll Virus (CLRV)

As part of a continuing series of observations on the occurrence of CLRV, which infects many woody hosts, the virus was found in leaves, fresh pollen or seeds of 43 out of 113 mature to over-mature common walnut trees *Juglans regia*, only 2 of which had yellow-brown ring foliar symptoms. Virus isolates from 30 widely-scattered walnut trees in Britain had few, if any, antigenic determinants not held in common with one another and with a walnut isolate from Italy (provided by Dr. A. Quacquarelli, Instituto di Patologia Vegetale, Bari).

The incidence of infection roughly paralleled the frequency of walnut, it being less (3/43) in northern England/Scotland, where walnut is scarce, than in southern England (40/70). However, because CLRV is seed-transmitted to 4–6 per cent of saplings, it seems that its greater occurrence in southern England might be attributed to a second method of transmission. CLRV was detected in 40 out of 1146 (4 per cent) walnut saplings tested from commercial nurseries and in 18 out of 300 (6 per cent) seedlings grown in methyl bromide sterilised soil.

In Britain, CLRV typically infects *J. regia* without changing leaf colour and shape. Nonetheless, infected seedlings (from virus-carrying seed) were found to be less vigorous than their uninfected counterparts, so probably necessitating prolonged periods of propagation. Measurements made on individually indexed *J. regia* seedlings 3-years old indicated that CLRV significantly decreased height growth from 144 to 109 cm and mean diameters from 17·5 to 15·6 mm.

CLRV was associated with an uncharacterized (carla)⁺ virus having rod-shaped particles with a modal length of 650 nm in *Sambucus racemosa* L. growing near Helsinki. Affected plants had chlorotic blotch ringspot symptoms in 'spring' foliage, but yellow ring and line patterns developed in leaves produced in summer and autumn. Whereas the antigenic determinants of CLRV isolated from Finnish *S. racemosa* were similar to those of an isolate from *S. racemosa* growing in East Germany (material supplied by Dr. J. Richter, Institut fur Phytopathogie, Aschersleben), both were distinct from isolates of CLRV from *S. nigra* L. and *S. ebulus* L (Plate 15).

Poplar Mosaic Virus (PMV)
Using a serological method reliably detecting 8 ng purified virus (enzyme-linked immune sorbent assay: ELISA) and infectivity assays (with a similar sensitivity), leaves from a range of poplar clones have been tested. Usually, similar results were obtained, but, late in the season, PMV was detected serologically but not when testing infectivity.

In one clone (Robusta), maximal concentrations of PMV occurred in leaves with symptoms. These appeared first in leaves produced in July, before subsequently appearing in younger and older foliage. Typically, virus concentrations were greatest in foliage occurring in the middle section of the current year's shoot, and least in foliage at the base of these shoots which expanded in cool weather (figure 31). The direct relation between symptom 'severity' and PMV concentration was confirmed by using 2 other *Populus x euramericana* clones (Lons and Regenerata). However, when specimens of Regenerata from a different

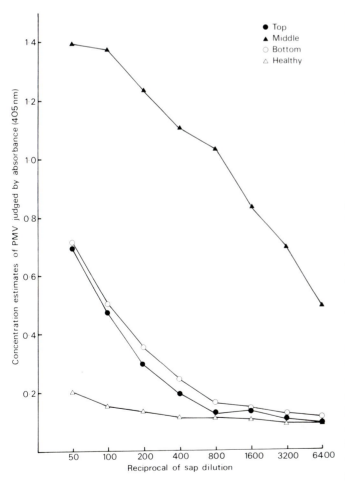

Figure 31 Concentrations of poplar mosaic virus (PMV), judged by absorbance (405 nm), and found in young (●), middle (△) and old (○) leaves in infected poplars.

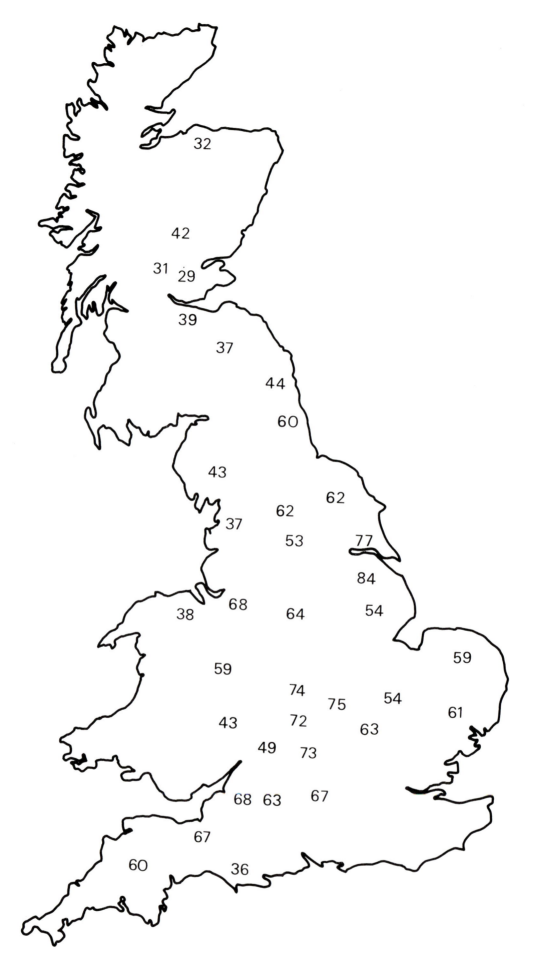

Figure 32   Estimates of the occurrence (mean % per county) of ash dieback in the UK.

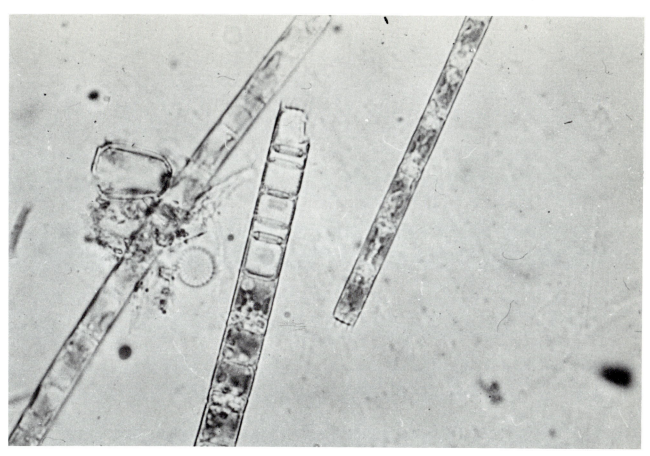

Plate 27   Threads of Melosira *species (between 5 and 9 μm wide)—September 1974.*

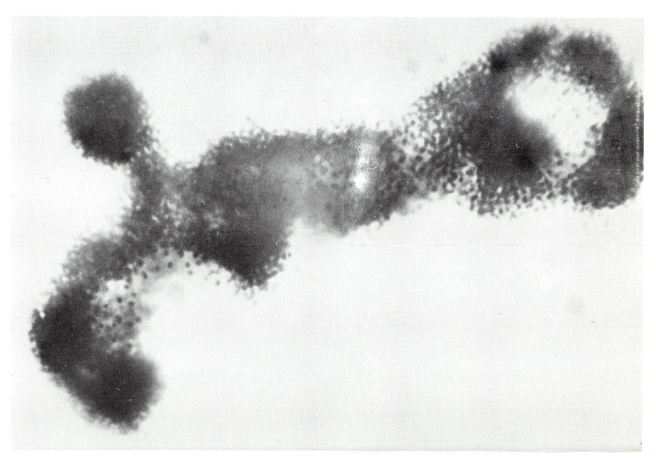

Plate 28   Microcystis aeruginosa Kutz., colony ca 400 μm across (August 1978).

Plate 29   Streaks of Microcystis *accumulating near the south shore of Loch Leven—aerial photograph taken from 1,000 feet—August 1978.*

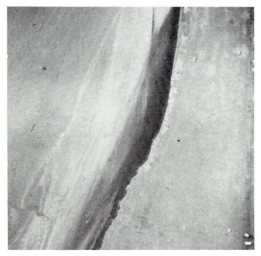

Plate 30   Microcystis *aggregations in the west bay of Loch Leven—false colour aerial photograph taken from 1,000 feet— August 1978—the algae show pink.*

(c) *Enclosures:*

Plate 31   (i)   *Experimental enclosures for algal-nutrient and phytoplankton-zooplankton interaction studies.*

Plate 32   (ii)   *Installation of large experimental enclosures for grazing studies.*
Photos: A E Bailey-Watts

plantation were tested, virus concentrations measured by ELISA and symptom severity were not associated. In contrast, symptoms were related to amounts of virus deduced from infectivity tests. It seems that a new isolate of PMV may have been detected, it being serologically distinct.

Arabis Mosaic Virus (AMV)

In the UK, where ash 'dieback' (plate 23) is prevalent (figure 32), particularly in industrial regions of Tyneside, the Midlands, and Humberside, AMV is found infecting *Fraxinus excelsior* L. However, when inoculated to ash, AMV seems capable of stimulating only slight foliar blemishes, which are partial and erratic. The occurrence of ash dieback in regions of Scotland, where soils have been found to be without the nematode *Xiphinema diversicaudatum* (Micol.) which transmits AMV, suggests that AMV is not the cause of the dieback.

J.I. Cooper

WOODLAND REGENERATION ON RED DEER RANGE

WOODLAND REGENERATION ON RED DEER RANGE

After centuries of felling and burning, only isolated patches of self-sown woodland remain in the Scottish uplands. However, the occurrence of saplings of Scots pine *Pinus sylvestris*, birch *Betula pendula* and *B. pubescens*, rowan *Sorbus aucuparia* and juniper *Juniperus communis*, which can be locally common, suggests that viable seed are often plentiful and that sites favouring germination exist. However, few saplings grow to maturity because of the depredations attributed to browsing red deer *Cervus elaphus* and sheep. Hence, many of the surviving stands are in danger of even further depletion because of the failure of the trees to replace themselves with seed-producing progeny.

The liability to browsing and the chances of self-sown and planted saplings growing taller than 30 cm were studied in the western Cairngorms, where red deer were effectively the only large herbivore. Local variations in deer densities were estimated from counts of faeces at designated sites scattered between 400 m and 900 m above sea level. On average, there was one deer per hectare. From May to October, deer ranged over the entire study area, with no appreciable difference in mean density at sites above and below 600 m. In winter, however, they concentrated on low ground which carried 5 times as many deer per hectare as sites above 600 m.

Any sapling that became established below 600 m had little chance of escaping damage or death after it grew taller than the surrounding vegetation. Whereas pines seemed equally liable to browsing throughout the year, similarly evergreen junipers, which seemed less palatable, were mainly browsed at times of food scarcity in winter. While inconspicuous during winter, deciduous species were browsed only sporadically, but, after budbreak, birch and rowan were immediately and continuously browsed intensively. Over 4 years, 18, 26, 36 and 80 per cent of unprotected self-sown saplings of birch, rowan, juniper and pine were killed, pines seeming to be unable to tolerate damage as shown in experiments assessing the effects of clipping.

Discounting effects attributable to the height of adjoining vegetation, damage to pine saplings decreased progressively with increasing altitude (figure 33). Like those of pine, plantings of other species, birch and juniper, were also browsed less at high altitude (figure 34). Saplings fared better at high altitude because amounts of browsing lessened during winter when deer migrated to low ground. At this time, November–April, pines on low ground were at risk, particularly when the surrounding vegetation was short. Between May and October, the amounts of browsing were independent of altitude.

It seems that wherever there is a large stock of free-ranging deer, the regeneration of tall woody plants is prevented on unfenced ground at low altitudes. However, given suitable seed beds and an adequate supply of seeds, regeneration might succeed at, or above, 600 m. Scrub at 530–590 m supported this conclusion. The frequency of the different size-classes of pine in relict pine-juniper scrub near 600 m indicates a distri-

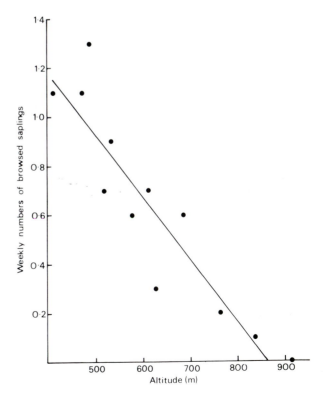

Figure 33 Numbers of planted Scots pine saplings browsed weekly at different altitudes in the western Cairngorms in the period November/April.

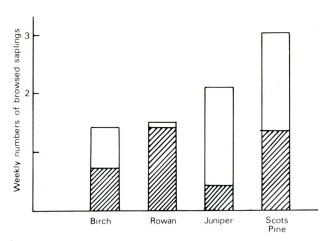

Figure 34   Incidence of browsing on saplings of birch, rowan, juniper and pine saplings planted at 800 m (□), or 400 m (□).

bution skewed towards the younger classes. Some 40 per cent of the 225 pines ha⁻¹ in the scrub were saplings aged less than 30 years and these averaged 41 cm in height, standing 22 cm taller than the ground vegetation —an unusual distribution in Scottish pinewoods, which mostly contain a predominance of large trees aged 100 years or more (Steven and Carlisle 1959). However, nearby saplings of the same age, but 300 m lower in altitude, were a mere 32 cm, being repeatedly browsed at, or below, the level of the surrounding vegetation.

G.R. Miller, J.W. Kinnaird and R.P. Cummins

Reference
Steven, H.M. & Carlisle, A. (1959). *The native pinewoods of Scotland*. Edinburgh: Oliver and Boyd.

TREE SELECTION STUDIES FOR THE REVEGETATION OF COAL WASTE

(This research is supported by the Department of the Environment)

The reclamation of coal waste is costly and not always successful. Many factors limit plant growth, including severe nutrient deficiencies, extreme pH, cementation and compaction, the occurrence of toxic substances and fluctuating amounts of 'soil' moisture. Current reclamation methods seek to ameliorate these conditions before planting, but problems can arise at later stages when the alleviation of adverse pHs and nutrient deficiencies may be more difficult to implement and increasingly expensive.

Despite difficulties encountered during reclamation, most sites eventually recolonise naturally as the surface spoil weathers and becomes more amenable to plant growth. By collecting cuttings and seeds of trees from these sites, it is hoped to obtain plants that are better able to tolerate reclamation sites than normal nursery stocks which are usually selected for their growth in favourable and fertile locations. Evolution of tolerance

to acidity, low pH, toxic substances (heavy metals) and deficient nutrient regimes have been detected in herbaceous plants, and Mason and Pelham (1976) found that some populations of *Betula* spp. grew conspicuously more than others when supplied with few nutrients.

Clonal stocks are being built up from cuttings obtained from *Betula pendula*, *B. pubescens*, *Alnus glutinosa* and *A. incana*, and a series of trials has been started. In the first of these trials, a pot experiment, clones of *Betula* spp. from coal spoil plus a control series from a commercial nursery were planted in June 1978 in (i) 2 types of coal spoil and (ii) compost; all plants were inoculated with mycorrhizal fungi of birch wood origin. Growth of all clones declined in coal spoil but continued in compost, with no significant difference in growth rates between clones. Interesting fruiting bodies of mycorrhizal fungi were found on coal spoil, these being associated with a greater incidence of mycorrhiza in spoil than in compost. So far, the benefits of using plant material of coal waste origin are not large, but this result is to be expected; until now there has been 'unconscious' selection of fast-growing, 'easy-to-root' clones which produce experimental material rapidly, a trait running counter to the evidence that ecotypes of grasses that are adapted to small amounts of nutrients are also slow-growing (Jowett 1959, Gemmell 1977). Until now, measurements have been restricted to shoot growth, so possibly overlooking significant effects on root growth which may be reflected in future enhanced shoot growth.

There is strong evidence from the USA demonstrating the importance of having the appropriate mycorrhiza for tree establishment on difficult sites (Marx 1975). Experiments testing the "effectiveness" of isolates of sheathing mycorrhizal fungi, obtained from coal spoil, are being structured.

J. Wilson

References
Gemmell, R.P. (1977). *Colonisation of industrial wasteland*. London: Arnold.
Jowett, D. (1959). *Some aspects of the genecology of resistance to heavy metal toxicity in the genus Agrostis*. PhD. Thesis, University of Wales.
Marx, D.H. (1975). Mycorrhiza and establishment of trees on strip-mined land. *Ohio J. Sci.*, **75**, 288–297.
Mason, P.A. & Pelham, J. (1976). Genetic factors affecting the response of trees to mineral nutrients. In: *Tree physiology and yield improvement*, ed. by M.G.R. Cannell & F.T. Last. London: Academic Press.

NUTRIENT CYCLING AND GROWTH IN RELATION TO AGE AND LIFE CYCLE STRATEGY IN *LYCOPODIUM ANNOTINUM* L.

The severity of the tundra climate is well-known (Lewis and Callaghan 1976), but edaphic conditions are also more unfavourable for plant growth than those of other regions. Tundra soils tend to be 'primitive' and decomposition rates are slow (Holding *et al.* 1974) and, as a result, there is a paucity of mineral nutrients available for plants. In order to be successful, tundra plants must, therefore, conserve essential mineral nutrients in addition to fixing, storing and utilizing solar energy efficiently.

Lycopodium annotinum, which is a successful tundra species and is widespread in the boreal forest floor vegetation of Fenno-scandia, was studied in the vicinity of the Abisko Research Station (which kindly provided experimental facilities) in Swedish Lapland. The long, severe winters and short, but favourable, growing seasons at Abisko produce innate markers of annual growth in plants of *L. annotinum* which enable the ages of segments of plants to be determined. From these 'historical' records, age-related patterns of growth could be determined (Callaghan and Collins 1976). Having constructed age-related life cycles, physiological processes could be interpreted in relation to developmental patterns (Callaghan *et al.* 1978). For example, it was possible to examine nutrient cycling within plants.

Life cycles

Like other clubmosses, *L. annotinum* has a life cycle with 2 distinct generations: a thallose *gametophyte* generation which persists underground for many years, reproduces sexually and gives rise to a leafy *sporophyte* generation which reproduces asexually. Only the leafy sporophyte generation was investigated, as the population at Abisko appeared to be maintained entirely by the vegetative reproduction of sporophyte plants. The leafy plant showed a 'modular' construction in that it consisted of annually-produced segments, the latter forming branches, some of which developed from 'horizontal segments' that creep through forest floor vegetation and develop roots (figure 35). As new segments are initiated at branch apices, older segments die. Horizontal branches can extend 7 cm per year. Other segments ('vertical segments') form aerial branches which support the third type of segment, a spore-producing strobilus.

Horizontal segments are generally 5 times heavier than vertical segments and strobili (figure 36). They take 8 years to mature, senescence being rapid after 21 years. In contrast, vertical segments develop quickly, achieving maximum dry weight after about 4 years when senescence begins (figure 36). Strobili are annual structures, initiation, development and spore release all occurring in one year, at the end of which senescence occurs and dry weight is lost exponentially (figure 36).

Photosynthesis is active in young horizontal and vertical segments, but strobili are only slightly photosynthetic in their first year. Although many of the segments in a plant—even actively-growing segments— are non-photosynthetic, translocation is extremely efficient and takes place between all segments while concentrations of translocated ^{14}C increase in senescing structures. Rates of translocation of ^{14}C in the field exceed 0·67 m per minute and changes in nutrient concentrations within aerial branches can only be explained by translocation.

Nutrient concentrations

The main sources of energy for respiration are soluble carbohydrates and fats. Soluble carbohydrate concentrations increase initially as the life span of the segments decreases from horizontal segments, through vertical segments to strobili (figure 37). The annual strobili show exceptionally large concentrations of over 50 per cent of the total dry weight. However, as the

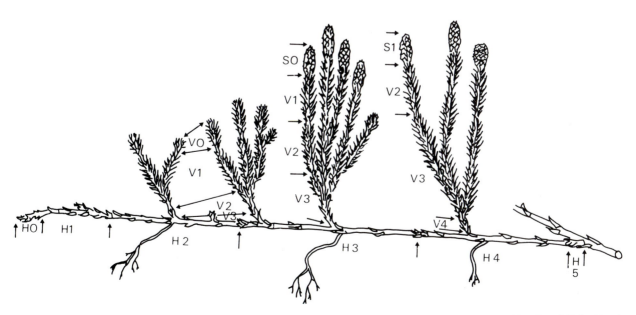

Figure 35 Growth habit of Lycopodium annotinum *L. Horizontal segments are marked H, vertical segments, V and strobili, S. Numbers refer to age in years of the different segments. Arrows denote morphological markers separating growth in successive seasons.*

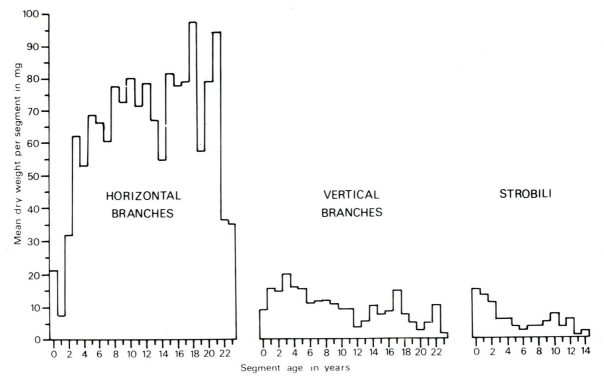

Figure 36   *Changing dry weights of ageing segments of (i) horizontal and (ii) vertical branches and (iii) strobili of* Lycopodium annotinum.

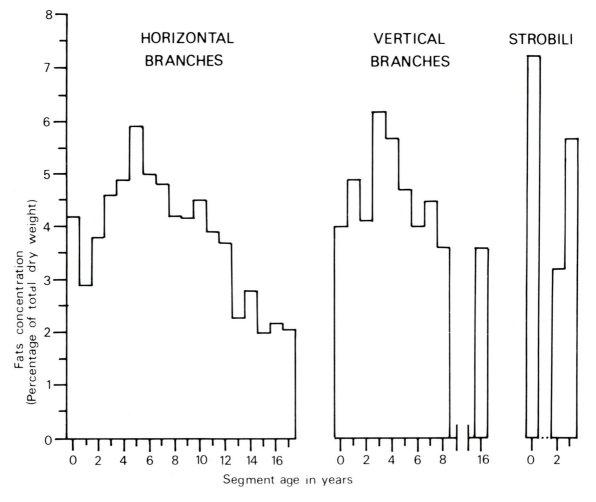

Figure 37   *Changes in concentrations of fats in ageing segments of (i) horizontal and (ii) vertical branches and (iii) strobili of* Lycopodium annotinum.

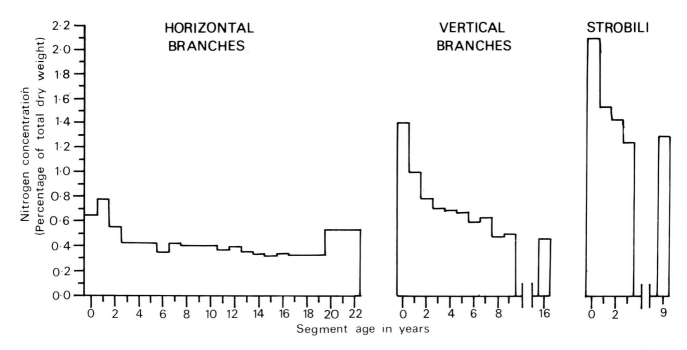

Figure 38 *Changes in nitrogen concentrations in ageing segments of (i) horizontal and (ii) vertical branches and (iii) strobili of* Lycopodium annotinum.

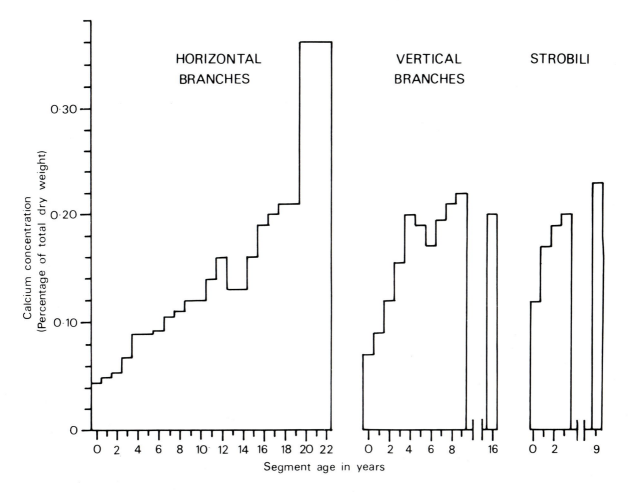

Figure 39 *Changes of calcium concentrations in ageing segments of (i) horizontal and (ii) vertical branches and (iii) strobili of* Lycopodium annotinum.

initial concentration increases, the rate of loss also increases so that concentrations decrease by a factor of 5 in 2 years in strobili, over 8 years in vertical segments, and over 19 years in horizontal segments. This decrease suggests that soluble carbohydrates are providing energy where physiological activity is greatest. Fats, in contrast, occur in similar concentrations in all 3 types of segment, and there appears to be a storage capacity associated with mature horizontal and vertical segments (figure 37) that have stopped photosynthesising.

Concentrations of nitrogen, an essential component of living matter, follow similar trends to those of soluble carbohydrates (figure 38). Increased nitrogen concentrations in senescing horizontal segments (figure 38) may be associated with the release of nitrogen due to the activity of microbial decomposers. In general, the range of nitrogen concentrations of L. annotinum from Abisko is small when compared with other species (even upland species of Britain) which range from 1 to 3 per cent.

Amounts of phosphorus, essential for trapping and releasing energy via its bonds with adenosine, change in the same way as those of nitrogen, with large concentrations (0·3 per cent) associated with the energy intensive process of spore production in the strobilus. Unlike that of nitrogen, the range of phosphorus concentrations in L. annotinum is comparable to that of other plant species.

Potassium is associated with osmotic processes, protein synthesis and membrane permeability. Again, concentrations decrease exponentially with increasing age. However, concentrations found in healthy, actively-growing horizontal and vertical segments range from 0·4 to 0·7 per cent. These amounts are extremely small when compared with ranges for 15 other species which, even when deficient in potassium, had 0·8 to 1·2 per cent (Evans and Sorger 1966).

Magnesium is an activator of many enzyme systems and is an important component of chlorophyll molecules. However, its concentrations do not appear to be related to photosynthetic potential in L. annotinum, as maximum concentrations occur in strobili after spore release. Concentrations in horizontal segments do not vary greatly over their 23-year life span.

Changes in calcium concentrations, unlike those of other elements, increase with age (figure 39). Calcium is an important component of cell walls, where it occurs as the insoluble salt of pectic acid which acts as a cementing compound, keeping cell walls rigid. It is a non-mobile element and remains as tissues age, whereas other elements are removed via translocation. Concentrations of calcium therefore increase as concentrations of other elements decrease.

The growth strategy and patterns of nutrient allocation in L. annotinum suggest an efficient system of energy and nutrient utilization. Mineral elements absorbed by roots attached to mature horizontal segments are translocated to actively-growing horizontal and vertical segments and strobili. Some elements, such as nitrogen and potassium, are only needed in very small concentrations—presumably because they are efficiently metabolised—and all elements except calcium (generally abundant in soil) are remobilised and recycled within the plant as segments age. Indeed, elements are only lost from the system (i) at spore release, which accounts for a very small fraction, and (ii) during the decomposition of old segments which, in any event, are depleted before the onset of decomposition.

Translocation of ^{14}C to senescing segments may, perhaps, represent the energy input necessary for the process of nutrient remobilization and translocation to younger segments. This efficient conservation of nutrients is important in tundra habitats where nutrient availability is a major factor limiting the growth of L. annotinum.

T.V. Callaghan

References
Callaghan, T.V. & Collins, N.J. (1976). Strategies of growth and population dynamics of tundra plants. I. Introduction. Oikos.
Callaghan, T.V., Collins, N.J. & Callaghan, C.H. (1978). Photosynthesis, growth and reproduction of Hylocomium splendens and Polytrichum commune in Swedish Lapland. Strategies of growth and population dynamics of tundra plants, 4. Oikos, 31, 73–88.
Evans, H.J. & Sorger, C.J. (1966). Role of mineral elements with emphasis on the univalent cations. A. Rev. Pl. Physiol., 17, 47–76.
Holding, A.J., Heal, O.W., Maclean, S.F.J.R. & Flanagan, P.W. (1974). Soil organisms and decomposition in tundra. Stockholm: IBP Tundra Biome Steering Committee.
Lewis, M.C. & Callaghan, T.V. (1976). Tundra. In: Vegetation and the atmosphere, ed. by J.L. Monteith, Vol. 2, 399–433. London: Academic Press.

GENECOLOGICAL STUDIES IN SPHAGNUM

Many species of Sphagnum are superficially very similar, yet, on closer examination, a wide range of variation can be detected within each of them. This variation may be directly attributable to the genetic composition of individual plants or populations, or it may represent the response of plants to differing combinations of environmental conditions. For example, submerged plants often have more lax growth forms than those in terrestrial habitats, or species which are normally distinctively coloured may become green when growing in shade. To assess the limits of variation, an attempt is being made to judge the genetic environmental influences on variation in Sphagnum pulchrum and S. flexuosum (sensu lato) which were chosen for different reasons.

S. pulchrum has a disjunctive distribution in Britain, being found mainly in low altitude, oceanic, raised and blanket mires of west Scotland and north-west England. Small quantities are found in west central Wales, but the species is present extensively in the valley mires of the Isle of Purbeck (Dorset). Are these populations composed of distinctive individuals and, if so, are the differences attributable to different climatic, hydrological or chemical conditions existing at the different sites? If they are not, do they represent the early stages of divergence in the formation of new species from a common ancestor?

S. flexuosum (=*S. recurvum*), unlike *S. pulchrum*, is widespread, occurring in a range of mire types with 3 morphological variants which continental taxonomists recognise as distinct species, namely *S. flexuosum* S.S., *S. fallax* and *S. angustifolium*, but which British bryologists treat as varieties. Each of these variants is considered to grow in a slightly different habitat, though their ranges overlap. In deciding whether or not they are distinctive species, it is important to know the extent to which the differences are genetically controlled and how much these differences reflect responses (phenotypic) to differing habitats. The extent of reproductive isolation is also of some interest.

Three approaches have been adopted to resolve these problems.

1. Samples are being collected from a large number of sites, and different parts of sites, throughout Britain in order to obtain a reasonable estimate of the total range of variation of 36 morphological characters, e.g. length and breadth of stem and branch leaves, porosity of leaf hyaline cell, and plant colour. Later, it is hoped to compare British material with that from the Continental European parts of their natural ranges.

2. Samples from a number of sites are being cultured in conditions with different water tables, light intensities and chemical constituents. Additionally, the responses of different plants from the same locality are being examined.

3. Electrophoresis is being used to discriminate between different genotypes, both between and within populations, in order to determine if individual populations have arisen as single clones and if the level of genetic variation is greater between than within populations.

Most of the samples of *S. flexuosum* so far examined, mainly from the north, south, south-west and west midlands of England, south Scotland and central Wales, correspond, in terms of stem and leaf shape, with the variant *fallax*, the variant *angustifolium* being the least common. From a preliminary examination, it appears that several characters considered as useful in separating *fallax* from *flexuosum* are not infallible. For example, the 30 per cent or more plants with sharply acute stem leaves so far examined lacked fibrils, though the presence of fibrils in sharply acute stem leaves has been considered an important secondary diagnostic feature for the variant *fallax*. It also seems that measurements of leaf length and branch length are of little value as they are likely to be influenced by external conditions. The extent of this variation was indicated when cultures were grown in conditions testing different (i) water tables, (ii) nutrient concentrations and (iii) pH. In dilute solutions of nutrients, growth did not respond consistently to changes in either pH or water level. In contrast, raising the water table in more concentrated solutions encouraged the production of shorter branches. (Cover).

The presence of different genotypes both within and between populations has already been established by the results of relatively few electrophoretic tests identifying polymorphism in the acid phosphatase enzyme system.

R.E. Daniels

PATHWAY OF FLUORIDE IN A GRASSLAND ECOSYSTEM

Introduction

It is increasingly being realised that an understanding of the processes involved in the transfer of toxic substances in the biosphere is essential if we are to gain an insight into, and have the capacity to predict, possible long-term effects on the structure and functioning of ecosystems. The natural processes involved in ecosystem energetics and nutrient cycles are beginning to be understood (Perkins 1978) and the methods developed for analysing the functions of the different components can now be applied to the study of pollutants. Pollution from airborne fluoride is of special ecological interest (Perkins 1973), not only because of its phytotoxicity (Chang 1975) and the occurrence of fluorosis in animals grazing on contaminated herbage (Allcroft *et al.* 1965), but also as a pathway tracer. Data are now becoming available from a long-term study in Anglesey which should enable the distribution, transfer and accumulation of fluoride to be examined in an ungrazed soil-plant-detritus system which, in due course, it is hoped to develop as a predictive model.

Site and Methods

The study site is 10 m above sea level and situated 0.65 km north of an aluminium reduction plant established in 1970 near Holyhead in the north-west coast of the island of Anglesey. The herb-rich vegetation, dominated by the grasses *Dactylis glomerata*, *Festuca ovina*, *Agrostis tenuis* and *Poa annua*, grows on a sandy soil with a pH varying from 5.9 at the surface (0–2 cm) to 7.1 at a depth of 5 cm. Bulk density was 1.19 g cm^{-3} and loss-on-ignition varied from 51 per cent in the top 1 cm to only 4 per cent at 9–10 cm depth. The grassland was only very lightly grazed by small mammals and occasionally by rabbits, effects of which appeared negligible. Ten 6 cm diameter turf cores were removed to a depth of 11 cm per month for measurement of live and dead plant biomass and other analyses (Perkins et al. 1978). Fluoride analyses were made with an F ion-selective electrode following alkali fusion of dried and milled plant material or soil. Water soluble F^- in soil was determined in a 5:1 water:soil suspension. Measurements were made at the site over an 18-month period from October 1975 to April 1977.

Distribution of dry matter and fluoride
The mean annual above-ground biomass of living plants (426 ± 56 g m^{-2}) contained 107 ± 29 mg F^- m^{-2} The weighted mean concentration of F^- was 251 ± 34 μg g^{-1} with dicotyledonous plants having the largest concentration (363 μg g^{-1}), grasses having 188 μg g^{-1}, and other 'monocotyledons' being in between with 236 μg g^{-1} (table 1). Grasses comprised 50 per cent (213 ± 30 g m^{-2}) of the total biomass and contained 40 mg F^- m^{-2}, whereas dicotyledons at 32 per cent (134 ± 43 gm^{-2}) contained 50 mg F^- m^{-2} (45 per cent of the total F^-) ; the remaining monocotyledons contained only 19 mg F^- m^{-2} in their 18 per cent (79 ± 14 g m^{-2}) share of biomass. Concentration differences between groups of species may reflect differences in absorption and retention of F^-, and these differences could be important if differentially selected by grazing animals. Concentrations in plants from a 'control' site at Aberdesach on the Lleyn Peninsula contained <5 μg F^- g^{-1}.

The concentrations of F^- in, and on, leaves depend upon rates of deposition, rates and quantities of rainfall and the stage of plant maturity. During dry weather, concentrations can increase by dry deposition and decrease following washing by rain, but interpretation is complicated because the emissions come from a point source and amounts falling at the site depend upon wind direction. For example, during periods of wet weather, when the plume is over the site, large concentrations similar to those found experimentally by Less et al. (1975) can be detected on foliage. Short-term, day-to-day, variations have been reported for F^- concentrations on herbage in north-east England (Davison and Blackmore 1976).

Table 1 *Mean annual concentration of fluoride (micrograms F^- g$^-$ dry weight) in components of the grassland ecosystem.*

Live Vegetation	
grasses	188 ± 29[a]
monocotyledos	236 ± 50
dicotyledons	363 ± 41
roots	309 ± 22
Dead Vegetation	
grasses	549 ± 36
monocotyledons	541 ± 72
dicotyledons	591 ± 93
litter	611 ± 48
Soil [0 to 11 cm]	
water soluble	21[b]
bound	115[b]

Footnotes: [a]Standard error indicates variation between 12 monthly samples.
[b]Composite value

The mean biomass of roots was 726 ± 66 g m^{-2}, most roots being concentrated at the surface and few below 11 cm. The ratio of above-ground : below-ground biomass was 1:1.7, a smaller ratio than would be expected from a heavily grazed grassland. Per unit ground area roots contained 217 ± 22 mg F^- m^{-2}, twice the amount found in living foliage, which in terms of dry weight is equivalent to 309 ± 22 μg F^- g^{-1}.

The mean biomasses of recently dead material still attached to living plants (standing dead) and litter (surface accumulations of plant debris) were 602 ± 36 g and 423 ± 47 g m^{-2} respectively. Standing dead contained 332 ± 34 mg F^- m^{-2} and litter 247 ± 30 mg F^- m^{-2}, together totalling 579 ± 26 mg F^-. These are larger concentrations than were detected on living foliage and the increases may be attributed to increased absorption, and also to an apparent increase as foliage loses dry matter content on senescence. Leaching can, however, take place from standing dead and litter during periods of high rainfall. In experiments at the main site, up to $6 \cdot 8$ μg F^- g^{-1} per day has been leached from material exposed for periods of 30 days.

The dry weight of soil to a depth of 11 cm was 131×10^3 g m^{-2}, containing $22 \cdot 7 \times 10^3$ g organic matter. It contained $17 \cdot 74$ g F^- m^{-2} (the water soluble fraction being $2 \cdot 7$ g and the bound fraction $15 \cdot 0$ g). Amounts of water-soluble (figure 40) and bound F^- decreased with increasing depth. At an uncontaminated control site, soil had negligible amounts of F^-, the soluble fraction being less than $0 \cdot 2$ μg F^- g^{-1}, compared with 10–40 μg F^- g^{-1} near the reduction plant.

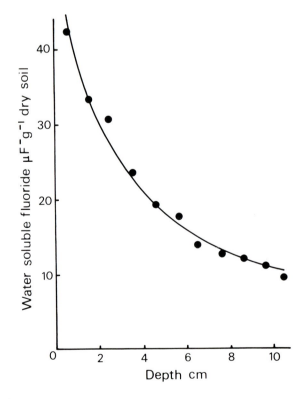

Figure 40 *Amounts of water soluble F^- found at different depths (d) in soil sampled from a herb-rich grassland near an aluminium reduction plant.*

Transfer of fluoride

Estimates of the disappearance of litter near the reduction plant, obtained by exposing nylon-net litter bags, indicated that k (the rate) $=2 \cdot 8$ mg g^{-1} per day. Assuming that conditions were unvarying, k can be used to estimate the flow of dead material through the decomposer system and thence primary production (Wiegert and McGinnis 1975; Perkins *et al.* 1978). In the event, the flow of dead material was found to be 1048 g m^{-2} yr^{-1}, containing 587 mg F^{-} (figure 2). Net primary above-ground production (*P*) (excluding material grazed) was estimated at 2221 g m^{-2} yr^{-1}

after applying a correction factor for ash of $\times 1 \cdot 79$. The F^{-} content of *P* was calculated, and possibly overestimated, by assuming a concentration of 273 μg F^{-} g^{-1}, the mean for living foliage during the season of active growth. Root production (340 g m^{-2} yr^{-1}) was estimated after calculating the 'turnover' rate (0·434) from biomass data.

Fluoride accumulation budget and pathway

Assuming negligible grazing, a budget for the grassland can be constructed (figure 41 and table 2). The flow of

PATHWAY MODELS OF

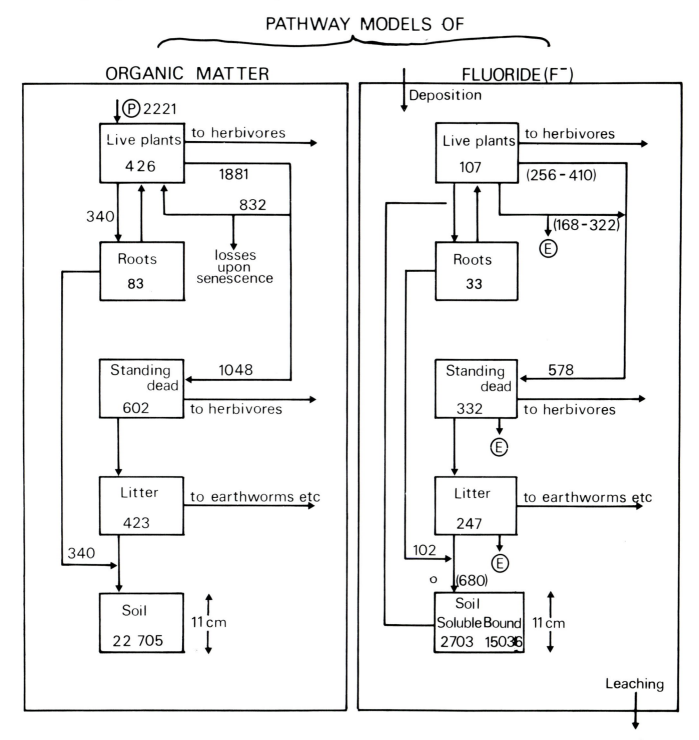

*Figure 41   Pathway models of organic matter and fluoride F^{-} through a herb-rich grassland near an aluminium reduction plant. Boxes give (i) the biomass of the different components and (ii) weights of soil organic matter (g m^{-2}). Numbers alongside arrows indicate the flows or transfers (g m^{-2} yr^{-1}) between the boxes. **P** = net plant production and **E** = fluoride eliminated by washing and leaching.*

Table 2 Fluoride budget for the grassland ecosystem at the end of 1976

	Fluoride in the system [gF^-m^{-2} to a depth of 11 cm.]
live plants	0.34
dead plants and litter	0.58
soil [water soluble]	2.70
[bound]	15.04

	Fluoride accumulated since 1970 [gF^-m^{-2} to a depth of 11 cm]
in plants [live and dead biomass]	0.90
in soil [less that in roots]	12.02
total accumulated	12.92

	Fluoride in transfer [$g\ F^-m^{-2}yr^{-1}$]
Through primary producers & de composers	0.68
Mean deposition rate over the 6 year period	2.15

F^- ($0 \cdot 68$ g F^- m^{-2} yr^{-1}) passes through 2177 g m^{-2} of living and dead plants containing $0 \cdot 90$ g F^- m^{-2}. Since 1970, it is estimated that total F^- in soil has increased by $12 \cdot 0$ g m^{-2} giving, with amounts in above- and below-ground biomass, a total of $12 \cdot 92$ g m^{-2}. This calculation suggests an average rate of deposition of $2 \cdot 2$ g m^{-2} yr^{-1} for the 6-year period since emissions commenced. Because the dead fraction contains 84 per cent of the F^- in the above-ground biomass, the decomposer system at this contaminated site is subject to the influences of large concentrations of pollutant F^-. As yet, the nature of these influences on the decomposingactivities of microbes and soil fauna is little understood. However, it is known that the augmented F^- concentrations affect plant growth, decreasing tillering and dry matter production.

D. F. Perkins, V. Jones and P. Neep

References
Allcroft, R., Burns, K.N. & Hebert, C.N. (1965). Fluorosis in cattle. 2. Development and alleviation: experimental studies. (*Ministry of Agriculture, Fisheries and Food, Animal Disease Survey Report 2*). London: HMSO.
Chang, C.W. (1975). Fluorides. In: *Responses of plants to air pollution*, ed. by J.B. Mudd & T.T. Kozlowski, 57–95. London: Academic Press.
Davison, A.W. & Blackmore, J. (1976). Factors determining fluoride accumulation in forage. In: *Effects of air pollutants on plants*, ed. by T.A. Mansfield, 17–30. London: Cambridge University Press.
Less, L.N., McGregor, A., Jones, L.H.P., Cowling, D.W. & Leafe, E.L. (1975). Fluorine uptake by grass from aluminium smelter fume. *Int. J. Environ. Stud.*, **7**, 153–160.
Perkins, D.F. (1973). Some effects of air pollution on plants, animals and soils. *Rep. Welsh Soils Discuss. Grp.*, **14**, 92–108.
Perkins, D.F. (1978). The distribution and transfer of energy and nutrients in the *Agrostis-Festuca* grassland ecosystem. In: *Production ecology of British moors and montane grasslands*, ed. by O.W. Heal & D.F. Perkins, 375–395. Berlin: Springer.
Perkins, D.F., Jones, V., Millar, R.O. & Neep, P. (1978). Primary production, mineral nutrients and litter decomposition in the grassland ecosystem. In: *Production ecology of British moors and montane grasslands*, ed. by O.W. Heal & D.F. Perkins, 304–331. Berlin: Springer.
Wiegert, R.G. & McGinnis, J.T. (1975). Annual production and disappearance of detritus on three South Carolina old fields. *Ecology*, **56**, 129–140.

Plant Community Ecology

ECOLOGICAL SURVEY OF BRITAIN

This survey has evolved from the application of numerical methods to the analyses of complex arrays of floristic and environmental data. It depends upon a close relation between environmental attributes read from maps and other records, and environmental observations made in the field. As these 2 sets of observations are strongly correlated, and as vegetation reflects habitat characteristics, it has become possible to predict the vegetation types of differing localities from environmental attributes read from maps.

Early studies in Grizedale Forest and the Lake District National Park (Bunce *et al.* 1975) led to the ecological survey of Cumbria (Bunce and Smith 1978). In this survey, data from physical, geological and human artefacts recorded on maps were analysed to produce 16 groups of km squares, termed land classes. These land classes have subsequently been characterised by field surveys of vegetation, soils, landscape and woodlands. The correlations between the classification derived solely from maps and the field surveys proved sufficiently high to enable valid predictions of observed characters to be made for regions where map information alone was available. Independent tests of the predictions proved that the method provided sufficiently accurate results for general trends to be established.

Although there is almost continuous variation, the land classes in Cumbria have recognisable characteristics and are readily identifiable in the field. Thus classes 7 and 8 are coastal and are predominantly mud or sand. Classes 6, 5 and 1 are lowland, with arable land predominating, and with many hedgerows and small copses. Class 4 is distinguished as being a lowland class but with a geology more usually associated with the uplands. Classes 2 and 3 are still lowland but are at higher altitudes than the previous classes, and hence tend to have more pasture and more open scenery. The remaining classes are upland in character, although classes 9–12 still have some relatively fertile land. Classes 13 and 14 are similar to those of 15 and 16, but, whereas the former are associated with the rounded slopes of the Pennines with much deep peat, the latter are present on the steep rocky slopes of the central Lake District and have mainly mineral soils.

The results of the Cumbria Survey have contributed to the structure plan drawn up by the County Council and have also been used in a variety of other projects, such as the comparison of valley systems for reservoirs, and an estimation of the county's potential for producing energy derived from plant material. Following the success of this exercise, the same principles are being applied to a survey of the UK. As for Cumbria, 'map' data are being assembled using a 1 km grid with 6,000 squares being characterised and used to establish 32 land classes—in contrast to the 16 classes which were found to be adequate for Cumbria. When this pre-liminary stage has been completed, 8 replicate squares of each of the land classes will be surveyed in greater detail, including vegetation assessments and descriptions of soil profiles. In due course, the ability to predict land characteristics will be tested objectively in anticipation of the requirements of an increasing number of agencies concerned with land classification, and the rational use of natural resources.

R.G.H. Bunce

References
Bunce, R.G.H., Morrell, S.K. & Stel, H.E. (1975). The application of multivariate analysis to regional survey. *J. environ. Manage.*, **3**, 151–166.
Bunce, R.G.H. & Smith, R.S. (1978). *An ecological survey of Cumbria*. Kendal: Cumbria County Council & Lake District Special Planning Board. (Structure Plan Working Paper 4).

PLANT SPECIES LISTS IN WOODLANDS

(This work was commissioned by the Nature Conservancy Council as part of its programme of research into nature conservation)

Many methods are commonly used to compile lists of plants growing in specified woodlands, ranging from informal wanderings to the formalised recording of species occurring in quadrats of a chosen size and distributed in a pre-determined manner. The characteristics of species lists compiled in these different ways, and the costs involved in their compilation, are being analysed as part of a wider investigation of woodland survey methods. Hales Wood, Essex, a mixed oak-ash-hornbeam woodland of 8 ha is one of the smallest in a series of woodlands being investigated. Five different types of survey, the most intensive being a series of 33 quadrats of 200 m² spaced at 50 m intervals throughout the wood, gave a total of 100 vascular plants, of which only 49 were recorded using all 5 methods. Numbers of species discovered by different methods ranged from 61 to 82 in the quadrat survey, with a mean of 71; similar differences occurred when the range of methods was applied to larger areas of woodland. Although the quadrat survey is the most expensive, this disadvantage is offset, at least in part, by the provision of a better base-line, including spatial distribution, against which temporal changes may be judged.

J.M. Sykes and A.D. Horrill

THE CHANGING GROUND FLORA IN UPLAND FOREST PLANTATIONS

(This work was commissioned by the Nature Conservancy Council as part of its programme of research into nature conservation)

To assess the alterations in upland Britain following extensive afforestation, vegetation in plantations in south-west Scotland, in north and south Wales and in north Yorkshire has been recorded over a 3-year period (1975–77) using a stratified-random programme of sampling. Observations were made of newly-planted areas, and on crops of different species and at different stages of development, attempt also being made to relate floristic differences with soil type. Additionally, it was possible to resurvey a forest in south Wales which had been recorded in 1944, shortly after planting, by an Oxford University team led by Mr. E. W. Jones (see Hill and Jones 1978).

Afforestation has an immediate impact on the abundance of ground flora species, with populations of *Calluna vulgaris* and *Vaccinium myrtillus* (dwarf shrubs) and *Molinia caerulea* and *Deschampsia flexuosa* (tall grasses) increasing at the expense of *Nardus stricta* and *Juncus squarrosus* and of *Trichophorum cespitosum* and *Narthecium ossifragum* in bog sites. Although these changes seem largely attributable to the cessation of grazing, other factors such as improved drainage and the application of fertilizers and herbicides must play a part. While bryophytes such as *Campylopus introflexus* and *Ceratodon purpureus* begin colonizing bare ground shortly after enclosure, their numbers and diversity increase greatly at canopy closure 15–20 years after planting, with *Lophocolea cuspidata* (growing mainly on dead wood) and *Plagiothecium undulatum* (on all substrata) being particularly abundant. Apparently, in western Britain, *P. curvifolium* owes its occurrence to afforestation, previously being virtually confined to deciduous woodlands in the east of the country.

As forest canopies become increasingly dense, with light intensities being minimal after 30–35 years, even bryophytes may be excluded, the effects of pine, larch and Douglas fir, which cast less shade, being less drastic than that of spruce and western hemlock. However, the complete suppression of ground species has been observed in thicket stage (15–20 years) larch.

Leaf litter, like shade, seems to restrict the growth of ground species, its effects sometimes contrasting with the luxuriant growth of bryophytes and ferns, sometimes found on litter-free tree stumps.

From a 'low' when plantations are 30–35 years old, the diversity of ground flora increases as forest crops approach maturity (say 50 years) (figure 42). The fern *Dryopteris dilatata* seems particularly well-adapted to this stage in forest rotations, while other species invade the spaces created by losses attributed to wind damage and other hazards.

Clear-felling, when "lop and top" is commonly left covering much of the ground, can severely disrupt the 'natural' development of ground flora. At this time, it seems that soil-borne spores and seeds play an import-

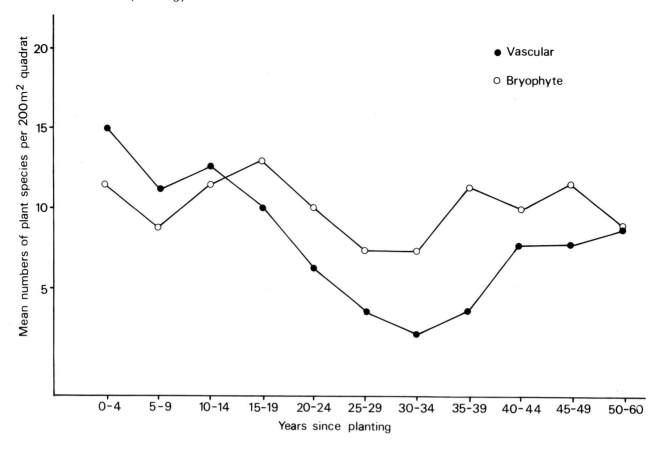

Figure 42   Numbers of species occurring in 200 m² quadrats under Sitka spruce related to the age of the crop.

ant part. Trials have shown that the re-establishment of ground vegetation can be partly attributed to seedlings from buried seeds that have survived through the crop rotation and partly to new immigrants. Thus, seeds of 'immobile' *Calluna vulgaris* and *Digitalis purpurea* survive *in situ*, whereas ferns and species such as *Epilobium angustifolium* and *Betula* spp. rely on wind-borne immigrants. 'Recovery' seems slower on peat than on mineral soils, possibly because fewer seeds survive.

To some extent, species diversity can be sustained by having contiguous blocks of forest planted at different times. Although we in Britain think of the herb-rich floia of Continental European boreal forests, it is perhaps more relevant to make comparisons with the upland plant assemblages before afforestation. In making these comparisons, the benefits of road verges, rides and streamsides should not be overlooked. Fifty-one species of vascular plants and bryophytes have been found in the verges of forest roads, compared with 26 species in the rides. In Dalby forest in north Yorkshire, the use of limestone for constructing forest roads greatly enhanced floristic diversity.

M.O. Hill and D.F. Evans

Reference
Hill, M.O. and Jones, E.W. (1978). Vegetation changes resulting from afforestation of rough grazings in Caeo Forest, south Wales. *J. Ecol.*, **66**, 433–456.

AUTECOLOGY OF THE MILITARY ORCHID

Orchis militaris, the military orchid, characteristically occurs in central Europe, extending eastwards into western and central Siberia and also occurring in mountainous parts of the Mediterranean region. It grows best in regions with warm springs and summers, being able to tolerate cold winters; it is rarely found where summers are hot and dry. Within the UK, it used to be found at scattered localities in south-east England along the Chiltern Hills, and in some of these areas was particularly abundant as, for example, around High Wycombe (Summerhayes 1951), but the number of sites is now very severely restricted (Perring and Farrell 1977).

On the Continent, its commonest habitat is chalk grassland, but, in England, it has generally been found at the margins or in the open parts of woods, amongst scrub, or occasionally on dry banks, but always on chalk. Some shelter seems desirable. The two 'major' English sites in Buckinghamshire and Suffolk are both wooded and protected by management agreements with the appropriate Naturalists' Trusts. The Buckinghamshire site, when first discovered in 1974 by J. E. Lousley, was relatively open chalk downland with few scattered hawthorn bushes, brambles and large yew trees. Subsequently, the site became completely covered with scrub, and, in the early 1960s, it was planted with beech and sycamore. The yew trees, which now shade large areas, are thought to be restricting the build-up of the population of *O. militaris*. The Suffolk site, located in 1954, is in a chalk pit in the middle of a coniferous plantation. Recently, encircling scrub was cleared, and, as a result, it is hoped that the population will increase.

To extend our knowledge of this species, detailed observations have been made of individuals in the Suffolk and Buckinghamshire populations, recording shoot and inflorescence emergence, and assessing numbers of seed-bearing capsules. Young shoots of *O. militaris* first make their appearance above ground in December, and grow slowly until May. During the last fortnight of this month, there is a period of rapid shoot elongation when the plants which are to flower increase from 5 to 15 cm in height. Summerhayes (1951) records well-grown plants up to 40 cm tall. At this stage, inflorescence is still enclosed in its sheath, and it is not until the week before flowering that it forces through its protective sheath, with the flowers opening within a few days (Plates 17, 18).

Flowering is acropetalous and lasts for about 3 weeks. Summerhayes (1951 p. 248) states that the aerial stem is first formed in the fourth year, and the flower-spike several years later, but, at the Gerandal Orchid Garden in Holland, flowering plants have been produced from seed within 4 years. Capsules take 7 weeks to develop and the seeds are ripe by early September.

Past records suggest that few flowers of an inflorescence (2–28 per cent) develop fertile fruits (Summerhayes 1951). In Suffolk, the proportion of fertile capsules, 3–10 per cent, assessed from counts made since 1976 on a random sample of plants, corresponds with the lower end of the limits given by Summerhayes. Whether this small percentage is a reflection of natural variation between populations, or is attributable to a decrease in seed production over the last 20 years, is not known, but it is hoped that continuing records will provide the answer.

Very little is known about the pollination of *O. militaris*, which may be subject to appreciable seasonal differences. Whereas *O. militaris* is visited, and presumably pollinated, by different types of bees, including bumble-bees, in Continental Europe (Summerhayes 1951), only Syrphid flies had previously been observed in the UK (see Lousley's field notes). In June 1978, the Syrphid *Baccha obscuripenis*, an aphid feeder found in woods and thickets, was observed visiting the Suffolk population, and, in addition, types of hoverfly, wasp, bumblebee, ant and sawfly. Despite the few species observed, it seems likely that many individuals visit flowers, as nearly every inflorescence had a resident spider.

It is widely believed that the sizes of orchid populations fluctuate wildly from year to year. The Suffolk population of *O. militaris* is no exception, numbers (total of

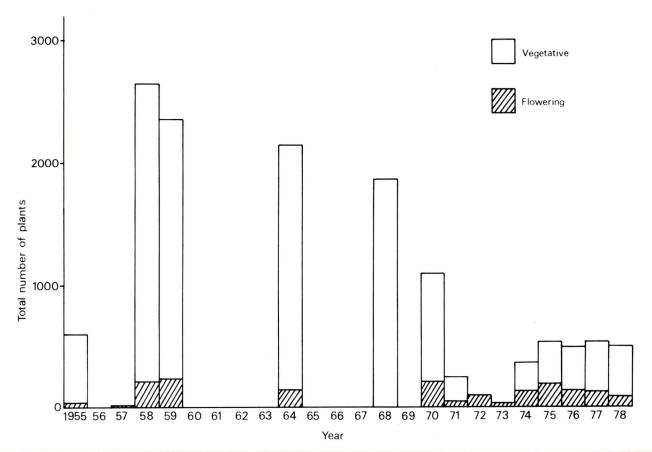

Figure 43 Population changes in Orchis militaris *at Rex Graham Nature Reserve 1955-78.*

vegetative and flowering) fluctuating from 2654 individuals in 1958 to only 252 in 1971. Over the past 5 years, the position of individual plants has been accurately plotted by a system of co-ordinates. The population has remained fairly stable since 1974, despite the extremely dry, hot conditions in 1976, and the mild, wet weather in 1978, with the proportion of flowering plants being greater than during the 1950s (see figure 43).

J.E. Lousley recorded the steady decline and final extinction between 1947 and 1965 of a colony in the Chilterns. As his sketches and field notes suggest that the demise of this particular colony was associated with dense shading, the effects of shade and its removal are being tested.

L. Farrell

References
Perring, F.H. and Farrell, L. (1977). *Vascular plants.* Lincoln : The Society for the Promotion of Nature Conservation. (British red data books : 1).
Summerhayes, V.S. (1951). *Wild orchids of Britain.* London : Collins. (The New Naturalist).

the growth of colonising grasses. When a stretch of 100 km of the Norfolk coast was surveyed by C. Turner, the 2 main dune-forming grasses, sea couch *Agropyron junceiforme* and sea lyme grass *Elymus arenarius*, were found to be absent from 17 locations in a 30 km section subject to erosion. Where they grew together—*A. junceiforme* sometimes occurred by itself—the mean percentage cover of *A. junceiforme* in 1977 was 3·6 times greater than that of *Elymus*. Flowering frequency in *Agropyron* was 81 per cent and in *Elymus* only 19 per cent. Generally, the viable seed output of *Agropyron* per m² is twice as large as that of *Elymus*, but, on the Norfolk coast, the output is nearly 8 times greater.

D.S. Ranwell

References
Dunn, J.N. (1972). *A general survey of Langstone Harbour with particular reference to the effects of sewage.* Portsmouth Polytechnic Report, 1–79.
Truscott, A.J. (1978). *Growth of Enteromorpha and salt marsh development in the Stour Estuary, Essex.* PhD. Thesis, University of East Anglia.
Turner, C. (1977). *Study of Agropyron junceiforme and Elymus arenarius.* ITE Sandwich Student report, 1–44.

SALT MARSHES AND SAND DUNES

Salt marshes
Extensive growths of green macro-algae occur on salt marshes in the Solent area and on the Essex coast. There is some evidence that these growths have developed relatively recently, possibly being linked with nutritional enrichment of estuarine and coastal waters (Dunn 1972). Studies in the Stour Estuary, Essex, completed recently by A.J. Truscott, suggest that, after being removed from the seaward edge of salt marsh by tides, mats of the algae *Enteromorpha* subsequently accumulate to landward on parts of salt marsh swards. Here, they have a detrimental effect on the growth of salt marsh plants, including glasswort *Salicornia europaea*, being responsible for the creation of bare areas susceptible to erosion. Mats of *Enteromorpha* decreased both the density and standing crop of *S. europaea*. Attempts to re-establish salt marsh vegetation using turf plugs on bare areas in the Beaulieu estuary, Hants., failed, possibly because of the increased erosive force of tidal water after the elimination of salt marsh vegetation.

Enteromorpha is only one of the increasing number of factors—land/sea level change, reclamation, pollution, etc.,—contributing to the destruction of salt marsh which performs a useful function in reducing the effect of waves on sea defences.

Sand dunes
Parts of the East Anglian coast are defended by the natural accumulation of sand dunes, accentuated by

INCIDENCE OF *PHRAGMITES* DIEBACK

Common reed *Phragmites australis* (Cav.) Trin. forms dominant communities in a variety of habitats. In the Norfolk Broads, it develops (i) dense marginal reed-swamp, and (ii) extensive reed marsh in adjoining low-lying areas. Yet, despite its versatility and tolerance, its occurrence in Broadland has decreased, particularly in pioneer reedswamps. The extent and rate of this change are being investigated.

A preliminary examination of aerial photographs, taken over the last 30 years, confirmed that areas of *Phragmites* had sometimes dramatically decreased. In 1949, 30 per cent or 11·4 ha of Hoveton Great Broad was reedswamp, the area decreasing to 5·4 ha and 0·07 ha in 1958 and 1974, respectively. (The 1974 aerial photograph data closely correspond to the 0·05 and 0·03 ha found during ground surveys made in 1977 and 1978.) On extending this study from selected broads to the whole of Broadland, it was found that *Phragmites* in most of the Bure and Ant Broads is similarly affected.

During the last 30 years, rates of sediment accumulation have increased six-fold, and *Phragmites* height growth was found to be inversely associated with depths of soft mud. However, other associations have also been detected, e.g. with nitrogen and phosphate.

L.A. Boorman and R.M. Fuller

ASPECTS OF PLANKTON ECOLOGY

In identifying the major causes of the different distributions (qualitative and quantitative, temporal and spatial) of fresh water plankton, both observational and experimental approaches have been used. While observations of Loch Leven have continued for a twelfth successive year, the project has been extended to include a limited number of other water bodies. Some of the additional work has been supported by contracts, but, more generally, it was designed to provide basic knowledge of how populations of plankton vary in different habitats. The extension provides a link between the intensive research on Loch Leven and the extensive regional synoptic surveys.

Phytoplankton in the shallow, eutrophic Loch Leven are thought to be strongly influenced by (a) types of grazing by different components, particularly Crustacea of the zooplankton community and (b) the external loads of phosphorus (Bailey-Watts 1978). Before 1971, phytoplankton populations were dominated by species with relatively small cells (e.g. *Synechococcus* n.sp, *Cyclotella pseudostelligera* Hustedt), but, with increasing populations of the filter-feeding herbivore *Daphnia hyalina* var. *lacustris* (Sars), phytoplankton species with much larger cells have dominated (figure 44 and plate 25). Furthermore, the seasonal pattern of changing cell size has altered. Whereas cell sizes before 1971 were maximal during winter, they are now largest during the summer months.

Cell size changes are important for many reasons, including their influence (i) on the nature and extent of light penetration into water columns (Bindloss 1976) and (ii) by virtue of differing ratios of surface area : volume, rates of nutrient uptake and energy gains and losses. Furthermore, the movement, and hence spatial distribution, of phytoplankton depends to some considerable extent upon the effects of water currents on cells of different sizes. In recent years, patterns of spatial heterogeneity have been particularly noticeable when, in calm weather, large, buoyant phytoplankton species, e.g. *Microcystis aeruginosa* Kutz and *Anabaena flos-aquae* emend Korarek and Ettl., have been abundant. These patterns of algal distribution contrast with the more usual and regular distribution occurring despite the consistently patchy distribution of zooplankton (figure 45). Because aspects of size determine rates of sedimentation and amounts of grazing, they must also influence the nature and extent of energy conversions within and between different trophic levels. (Plates 26-30)

Three types of experiments are investigating the nature of zooplankton grazing, each depending on natural assemblages of phyto- and zooplankton; experiments with single species are planned for the future. Short-term experiments, of 5 days duration, have been carried out at fortnightly intervals throughout 1978. In each, 3 treatments were tested : (1) untreated water from Loch Leven, (ii) water from Loch Leven from which the larger zooplankton elements were removed, and (iii) water from Loch Leven supplemented with zooplankton.

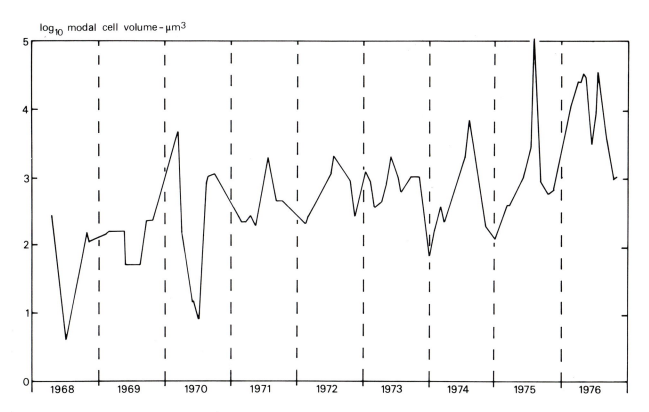

Figure 44 Fluctuations in the size (volume) of the dominant phytoplankton individuals in Loch Leven 1968-1976 (from Bailey-Watts & Kirika—in prep.)

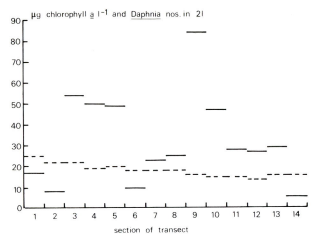

Figure 45   Concentrations of phytoplankton (as chlorophyll a —dotted bars) and Daphnia *(solid bars) in contiguous sections along a straight transect across the southern zone of Loch Leven-June 1977 (Data of George & Bailey-Watts using the tracking system of George 1976).*

Preliminary analyses suggest that the many changes in the numerical abundance, and sizes of phytoplankton species are controlled by numbers of zooplankton species and the balance of *Daphnia* to *Cyclops*, these influences being secondary, however, to those initiated by chemical factors, e.g. shortage of nutrients (Bailey-Watts 1976). (Plates 31, 32)

The other types of experiment involve the use of enclosures of 0·5 and 50 m³, the smaller enclosures being suitable for experiments of 2 or 3 weeks duration, whereas the large enclosures, based on the Lund tube design (Lack and Lund) 1972, facilitate continuous observations for periods of 2 or 3 months. So far, the gross physical, chemical and biological features of the larger enclosures have been characterized in comparison with the features of unenclosed bodies of fresh water.

For the future, experiments will be done to quantify and explain interactions between each of the major components of the plankton food chain, from dissolved nutrients *via* phytoplankton and herbivorous rotifers and crustacea, through carnivorous forms to juvenile fish (perch).

A.E. Bailey-Watts

References
Bailey-Watts, A.E. (1976). Planktonic diatoms and silica in Loch Leven, Kinross, Scotland: a one-month silica budget. *Freshwater Biol.*, **6**, 203–213.
Bailey-Watts, A.E. (1978). A nine-year study of the phytoplankton of the eutrophic and non-stratifying Loch Leven (Kinross, Scotland). *J. Ecol.*, **66**.
Bindloss, M.E. (1976). The light-climate of Loch Leven, a shallow Scottish lake, in relation to primary production by phytoplankton. *Freshwater Biol.*, **6**, 501–518.
George, D.G. (1976). A pumping system for collecting horizontal plankton samples and recording continuously sampling depth, water temperature, turbidity and *in vivo* chlorophyll. *Freshwater Biol.*, **6**, 413–419.
Lack, T.J. and Lund, J.W.G. (1972). Observations and experiments on the phytoplankton of Blelham Tarn, English Lake District. 1. The experimental tubes. *Freshwater Biol.*, **4**, 399–415.

Soil Science

RELEASE OF ELEMENTS FROM PARENT MATERIALS TO SOILS AND FRESH WATER

A collaborative programme of research has been initiated to study geochemical cycling, involving colleagues in the Institute of Geological Sciences (IGS), the Institute of Hydrology (IH) and the University College of North Wales (UCNW), in addition to the Institute of Terrestrial Ecology (ITE). The investigation is to be focused on the soils of 2 small instrumented catchments, one being at Plynlimon where the IH has gained considerable experience over many years, and the other probably in the Lake District, at a site with a contrasting parent material.

The collaborative programme is of 3 parts, which when integrated should enable mass transfers and nutrient budgets to be formulated:

1. Experimental weathering (UCNW): The main mineral 'species' and whole soil samples will be experimentally weathered using a variety of extractants/reagents at a range of pH, Eh and temperature conditions both in 'sterile' and biologically active conditions. The rates of release of a group of elements, including Ca, Mg, K, Fe, Al, Mn and P, will be measured.

2. Movement of elements (ITE): A series of laboratory leaching experiments and field-based lysimeter investigations will determine the movement of selected elements within and through soil profiles. In the former, soil cores and blocks will be leached/extracted with a variety of extractants in a range of environments, the within block measurements being examined with miniature porous ceramic cups and glass fibre bundles. The lysimeter studies will exploit ceramic cup soil solution samples and Ebermeyer-type lysimeters, samples of soil solutions being taken from within and below each horizon of the differing soil profiles. As in the experimental weathering section of the project, rates of movement will be determined.

3. A study of the relation between trace element contents of soil and bedrock on the one hand and in stream sediments and pan concentrates on the other (IGS). Using successional samples, it is hoped to characterize seasonal patterns of variation.

M. Hornung

TREES AND TOADSTOOLS (OF MYCORRHIZAL FUNGI)

Over the years, the numbers of toadstools (agaric sporophores) appearing in the autumn around trees in a collection of birch *Betula pendula* and *B. pubescens* planted near Edinburgh 7 years ago have progressively

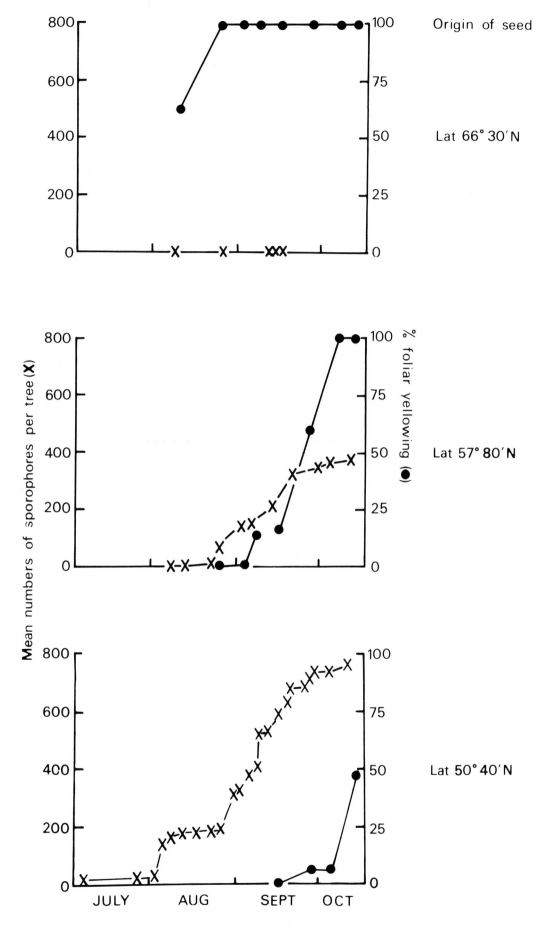

Figure 46   Seasonal leaf yellowing and numbers of sporophores
of mycorrhizal fungi association with Betula pendula trees grown
from seed collected at three latitudes.

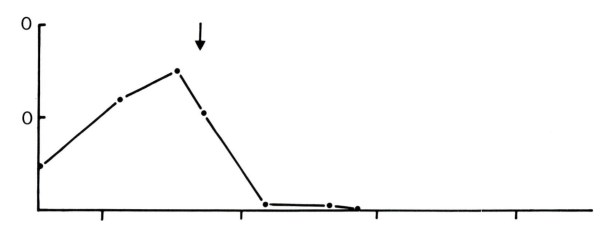

A   Defoliation on 22 August

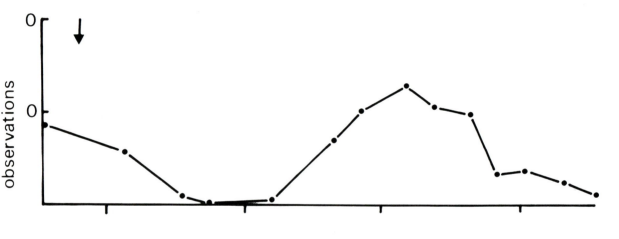

B   Defoliation on 25 July

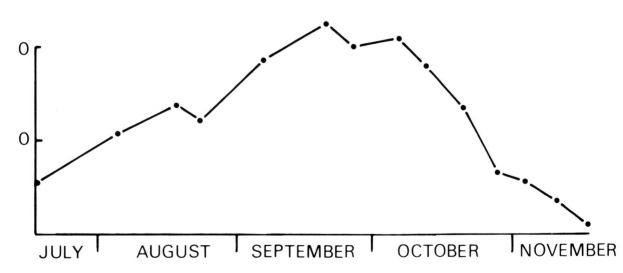

C   Controls naturally defoliating from October onwards

JULY   AUGUST   SEPTEMBER   OCTOBER   NOVEMBER

Figure 47   Effects in 1978 of defoliation on numbers of birch saplings, of 64, producing sporophores of mycorrhizal fungi in the intervals between successive observations.

increased. Whereas some seem to be scattered, others occur in recognisable circles (fairy rings) with clear evidence that diameters of different circles, which may be concentric, increase from one year to the next. Outstandingly, however, numbers of toadstools, all of fungi that can form mycorrhizas, differ according to the geographical origins of the different seedlots (provenances). Many more (800 per tree) were associated with trees derived from seed collected at a southerly latitude (50°41′N) than developed around trees from seed from latitude 66°30′N in the north of Sweden, namely 1–2 toadstools per tree per year. Interestingly, the northern trees lost their foliage by the end of August, whereas those from latitude 50°41′N did not start to yellow until mid-October (figure 46).

To investigate the possible dependence of toadstool production on the presence of active foliage on trees whose roots they colonize, 48 saplings of each of 2 clones of each birch species were planted at 3 m spacings on an agricultural site in 1975. In due course, their roots were colonized naturally by mycorrhizal fungi, with the first toadstools being observed in 1977. During 1978, many more developed, the first appearing in early July and the last in early November, the sequence, noted at intervals of 7–14 days, including *Inocybe petiginosa, Laccaria laccata, L. proxima, Peziza badia* and 2 *Hebeloma* species. During 1978, 3 treatments were tested: complete defoliation, using scissors (i) at the end of July; (ii) at the end of August; and (iii) a naturally defoliated control series that began to lose foliage in mid-October. By the end of July, 24 of each treatment group of 64 trees had produced toadstools; from August to September, numbers of the control trees with newly-produced sporophores increased to a maximum of 40 per fortnight, subsequently to decrease in late October with the onset of natural defoliation (figure 47): in total, 38, 54 and 55 of the control trees produced sporophores in August, September and October, respectively. Experimental defoliation in late August completely curtailed the subsequent appearance of toadstools, an effect duplicating the earlier response to treatment at the end of July. However, whereas terminal and axillary buds remained dormant after August defoliation, they flushed after the earlier treatment, with leaves becoming increasingly plentiful from September onwards. The appearance of leaves was paralleled by the resumption of toadstool production (figure 47), numbers of trees with newly-produced toadstools progressively increasing from 2 to 16 and 44 in August, September and October, respectively.

Clearly, the production of mycorrhizal toadstools is host-dependent, with the response to defoliation being nearly instantaneous. This relation may reflect changes attributable to the differing availability of current photosynthates, but the immediacy of the response suggests that the enlargement of toadstool primordia, if not their initiation, is controlled by another type of host factor, possibly involving growth regulatory substances.

J. Pelham, P.A. Mason and F.T. Last

MICROBIAL DECOMPOSITION OF LEAF LITTER OF TREES AND SHRUBS

Different aspects of the decomposer network were studied at Meathop Wood, an IBP site. To assess the relative contributions of fauna and micro-organisms, a field method was developed for studying the decomposition of litter by microbes in the absence of animals (Howard and Howard 1974). Weight loss, respiration, changes in C, H, N, and loss on ignition (LoI) of hazel *Corylus avellana* L., ash *Fraxinus excelsior* L., oak (chiefly intermediates between *Quercus petraea* (Mattuschka) Liebl, and *Q. robur* L.), and birch *Betula pendula* Roth. litter were each studied during 3 periods, each of 2 years: 1966–68, 1967–69, and 1968–70. Elm *Ulmus glabra* Huds. was studied in 1966–68, lime *Tilia cordata* Mill. in 1966–68 and 1967–69, and hawthorn *Crataegus monogyna* Jacq. in 1968–70. The analyses of the resulting comprehensive data set, which are now completed, presented many statistical problems (Howard and Howard 1974; in press a); in press b).

To compare weight losses in different species and in the same species in different years, it was necessary to find the best-fitting mathematical functions because sampling times differed. These functions reflect the form of the weight change curves, and, in fitting them, queries are raised about factors influencing the shape of the curves. For each set of observations, the following regressions were computed: asymptotic ($W = A + B r^t$), quadratic ($W = A + Bt + Ct^2$), exponential ($W = A.B^t$ or $\log W = a + bt$), and power function ($W = At^n$ or $\log W = a + n \log t$), where W is the percentage of the original dry weight remaining, t is the time in days, $a = \log A$ and $b = \log B$. Whereas other authors have used an exponential function to characterize weight losses, a better fit was usually obtained with asymptotic regressions (Howard and Howard 1974), litter losing weight rapidly during the first 20 days. Rates of litter decomposition within a species did not differ among the 3 experimental periods, with the notable exception of hazel. Thus, except for hazel in 1967/68, the data for the differing periods can be represented adequately by the asymptotic regressions of the pooled data for each species (figure 48). The regressions for litter of oak, hazel, birch, lime, and hawthorn are similar; they differ significantly from those of ash and elm whose litter decomposed more rapidly.

Although the N content of litter of oak, hazel, elm and lime fluctuated, often appreciably, changes in the gross composition of litter, analysed as C, H, N and LoI, did not show discernible trends with time. On the other hand, the C/N ratio clearly decreased with time.

Having measured the respiration of litter of oak, hazel, ash and hawthorn, it was found that the Q_{10} and Arrhenius models fitted the oak data equally well, and both fitted better than a simple linear model. For ash, hazel and hawthorn, the simple linear model usually fitted the data better than either the Q_{10} or Arrhenius models. Initially, respiration at 10°C was selected to give a simple representation of the progress of decomposition. However, it proved very difficult to fit an equation to the data because of the complex effects of moisture, the best-fitting model being a linear function of respiration on the percentage of remaining dry weight. Using this model, the predicted respiration curves showed, as expected, that litters with large respiration rates lost weight rapidly.

Most studies were carried out in Meathop Wood, with a mull humus layer. For comparison, the decomposition of birch and oak litter was also studied at a site on acidic rocks with soil having a mor humus layer (Bogle Crag Wood). Changes in weight loss, and total N and rates of respiration in birch litter differed consistently and significantly from those of oak. Other statistically significant effects were sometimes detected. The source of the litter had an influence on its moisture retention and pH; similarly weight losses, moisture contents, rates of respiration and pH were differently influenced by soil type at the 2 sites. Litter at the mor site, which had the greater rainfall, dried sooner after rain than litter at the mull site.

This work has indicated the need for a greater understanding of the physiology of mixed microbial populations. How do they respond to the different external factors controlling decomposition and how can these responses be predicted?

P.J.A. Howard and D.M. Howard

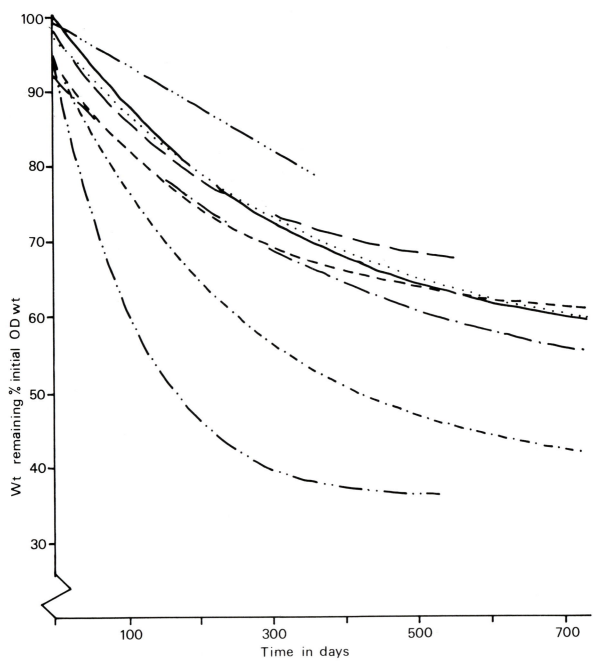

Figure 48   Asymptotic regressions of weight remaining on time of pooled data for each species :   elm ——··——, ash ——·—, hawthorn ——·——, hazel (1966/68 and 1968/70) ——·———. lime ——— ——— ———, birch ····, oak ———, hazel 1967/68) ——···——.

References
Howard, P.J.A. & Howard, D.M. (1974). Microbial decomposition of tree and shrub leaf litter. 1. Weight loss and chemical composition of decomposing litter. *Oikos*, **25**, 341–352.
Howard, P.J.A. & Howard, D.M. (in press). Microbial decomposition of tree and shrub leaf litter. 2. Respiration of decomposing litter. *Oikos*.
Howard, D.M. & Howard, P.J.A. (in press). Microbial decomposition of tree and shrub leaf litter. 3. Effect of source of litter and type of soil. *Oikos*.

FAUNA/MYCOFLORA RELATIONSHIPS

Studies of leaf litter decomposition have concentrated on separate analyses of microbes and fauna. Interactions between the 2 have received less attention, although it has been recognised that the ingestion of fungi by fauna may influence the balance and succession of microbes.

To determine the effects of invertebrate grazing on the competitive balance between fungi, 4 species of basidiomycetous fungi were selected for study after noting the common occurrence of their fruit bodies on Sitka spruce litter at Grizedale Forest, Cumbria. The

Table 1. Competitive balance between *Mycena* and *Marasmius*

Species	Starting ratios		
	75/25	50/50	25/75
Marasmius	93	78	51
Mycena	7	22	49

χ^2 test of departure from starting ratio
= 84.69 with 2 degrees of freedom (P <.001)

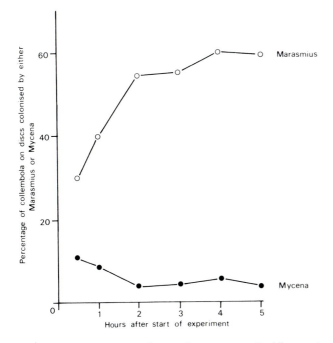

Figure 49 Collembolan preference between two Basidiomycete fungi.

collembolan *Onychiurus latus*, which preferentially ingests the mycelium of basidiomycetes and occurs in large numbers with the same spatial and seasonal distribution as the selected fungi, was chosen as the experimental animal.

To assess their competitive ability, the 4 selected fungi, after being grown separately on leaf litter, were mixed in different proportions before being inoculated to sterile spruce needles. Subsequently, needle colonisation by the morphologically-distinct fungi was measured. In different mixtures with *Mycena galopus*, *Marasmius androsaceus* colonised more needles than would be expected (Table 1). When confronted with mixtures of these fungi, *O. latus* preferentially selected *Marasmius* (figure 49). Thus, early results obtained in the laboratory suggest that *Marasmius* is both competitive and more heavily grazed than *Mycena*.

K. Newell

Data and information

BIOMETRICS

Routine advice on the design and analysis of experiments and surveys continues to be provided, but 3 projects with major inputs are highlighted.

1. *Analysis of Common Bird Census*
    (This work was commissioned by the Nature Conservancy Council as part of its programme of research into nature conservation)

The Common Bird Census was begun in 1961 by the British Trust for Ornithology to monitor population changes among the commoner British birds. The changes are measured by a breeding season census at a number of sample sites, repeated annually whenever possible. Originally, the data were analysed by contingency table theory. After some years, it became clear that this form of analysis was producing anomalous results; the variation in counts is much less than expected. The anomalies are explained to some extent by territorial behaviour; standard contingency table theory requires the number of birds in a woodland to have a probability distribution formed from the sum of variables distributed independently of each other; in a territorial situation, this is not the case, as the presence or absence of a bird depends on the number of other birds holding territories.

A mathematical formulation of the counts is further complicated by the possibility of serial correlation between the counts of the same sample area in successive years. The numerical analysis is made more difficult by the non-orthogonality of the data; some areas are sampled in some years and not in others.

A mathematical model incorporating territorial behaviour has been formulated and methods of analysis have been developed. Initial tests of the method using simulated and field data have given encouraging results.

M.D. Mountford

2. *A measurement of familial resemblance for ordinal data*

Experimental studies of the factors affecting the aggressive behaviour of red grouse have raised the problem of detecting and measuring similarity in aggressiveness of male birds from the same clutch. From observations on the interactions within a group of birds and by successive removal of individuals, a dominance hierarchy can be produced. The aggressiveness of a bird is measured by its position in the hierarchy—its dominance rank.

The problem of testing for familial resemblance using rank data is straightforward, but the measurement of the degree of resemblance is not. Methods developed for parametric measurements, such as the intra-class correlation coefficient, are not appropriate because they depend directly on the size and structure of the dominance hierarchy; comparisons between hierarchies could be biased.

A measurement of familial resemblance which overcomes the above problem has been devised and some of its other sampling properties have been studied. Comparison of the measurement with existing optimal methods in parametric situations has shown that it is also a relatively efficient detector of familial resemblance. The method has been used to compare familial resemblance in groups of birds obtained from different breeding programmes, and also to monitor familial in a fluctuating wild population.

P. Rothery

3. *Biological classification of rivers in Great Britain*

Statistical advice and analysis is being supplied by ITE for this project which is being run by the Freshwater Biological Association's River Laboratory at Wareham, under the sponsorship of NERC and the Central Water Planning Unit.

Geology and climate result in considerable variation in British river systems, ranging from stable chalk streams in southern England to mountain rivers subject to spate in the north and west. Differences in physical and chemical characteristics give rise to differences in biological communities of invertebrates, plants and fish, both along the course of individual rivers and between rivers. The aim of this project is to understand the basis of these differences and, in particular, to relate the invertebrate communities to major environmental features.

Forty-one key rivers have been selected, covering all river types found in Britain, and a co-operative programme of collection and identification of organisms from these rivers is being undertaken by the biologists of FBA and the Water Authorities. Multivariate clustering techniques are being applied to the resulting species lists so as to identify groupings of invertebrate species which will be used to form a biological classification of river zones.

Correlations will then be sought between the occurrence of different biological communities and physical and chemical factors at the sampling site, with a view to constructing predictive models relating the structure of biotic communities to environmental characteristics. These models will be tested and, if valid, they will be applied in drawing up a system to classify all river zones in Great Britain for which adequate data are available.

D. Moss

COMPUTING

The provision of interactive computing facilities to the members of other Subdivisions continues. This activity involves software support and the writing of specific computer programmes. Work is also continuing on overcoming the problems of inter-changing computing jobs, programmes and data bases between Research Stations. The inter-change of computer programmes written in BASIC and FORTRAN IV is increasing as a result of improvements in standardising both the major languages.

There are major problems of data-processing congestion at some Stations with large amounts of weekend and night-time computing. These problems will be alleviated by the installation of a PDP 11/34 at Merlewood and the upgrading of the PDP 11/10 at Monks Wood and Bangor. These enhancements will allow efficient access to the NERC computer network which should help considerably in providing standard computer facilities and data transfer between the major users in ITE, and will provide a better service through time-sharing facilities.

BIOLOGICAL RECORDS CENTRE DATA BANK

The Biological Records Centre (BRC) has held data on species distribution for many years and produced maps showing these spatial distributions. Until recently, the maps were produced using a modified typewriter, and there has been little opportunity to interrogate the data bank for other purposes. All the data previously held by BRC in computer-readable form have now been transferred to the Science Research Council's twin IBM 360/195 computer configuration at the Rutherford Laboratory. This centre houses the NERC Central Computing Group which is responsible for running and developing G-EXEC, a generalised data management system, originally created for geologists, but which has been found to be useful for other types of data banking. ITE has chosen to use the G-EXEC system for the Biological Records Centre data bank because it provides nearly all the facilities required and also opens up the possibility of associating ITE's data with those held by other NERC Institutes.

The data currently held in the computer consist of:

Vascular plants	1,350,000	records
Birds	285,000	records
Moths	175,000	records
Non-marine molluscs	140,000	records
Bumblebees	9,000	records

The total of nearly 2 million records is still increasing: data for the Geometrid moths will be added shortly. The vascular plant data are incomplete because the individual record cards (about 10 per cent of the records) were omitted when the data were originally transferred to computer storage in 1970. Work has started on editing and punching these individual record cards for inclusion next year. The omission of these latter records from the data bank is a serious drawback and prevents full analyses of the plant records.

The information held in each record consists of the following items: species and sub-species numbers, grid reference, vice-county, date, rarity (local), status (normally whether native or introduced), source (field, museum or literature), expert confirmation (where applicable), and data type. Not all these items are recorded for every species, or every scheme, but the minimum requirement is species number, grid reference (10 km), date (or date class), and data type. The last item describes where the original record can be found. Whenever possible, the vice-county is included as this enables a cross-check to be made on the grid reference. When the plant individual records are added, habitat and altitude will be stored in addition to the above items.

The plant individual records will be read from punched cards. Punching data from species list cards on to paper tape requires a great deal of sorting and pre-processing before the data are ready for use and is very slow and expensive. In recent years, we have experimented with the use of optically-read forms, where the presence of a species is recorded by a pencil mark in the appropriate box on the form. The use of these forms for the Lepidoptera scheme, using an OPSCAN reader, has considerably speeded up the input of data, and has resulted in maps being available more quickly. We shall use this method of input whenever possible in future.

Many requests for information from the data bank will simply require a list of records retrieved by applying limits to one or more of the fields comprising a record,

and this type of output can be obtained from any information retrieval system. However, one of the advantages of G-EXEC is that, in addition to the normal data management and retrieval facilities, it also provides programmes for data and information, statistical analysis, plotting and, it is planned, modelling. Since most of BRC's data are collected initially as part of a mapping scheme, the plotting package (G-PLOT) is particularly useful. An example of G-PLOT output is given in figure 50, which shows the distribution of *Salix caprea* (goat willow) in Wales. This map was generated by a single G-EXEC job, with the exception of the actual plotting process, which was done by the FR80 microfilm recorder. The successive stages in the G-EXEC job were: selecting the data for a species; rejecting records outside a given (rectangular) area; transforming the grid reference into mapping co-ordinates; generating the mapping symbols, and grid and annotations; and generating the outline (in this case a subset of the full outline of the British Isles).

G-PLOT will probably be used to supply customers with single copies of maps, and to provide compilers of atlases with preliminary maps for circulation to recorders, while the production of large batches of maps for publication will be done by the Experimental Cartography Unit (ECU). The first atlas to be produced jointly by BRC and ECU, the *Provisional atlas of the bryophytes of the British Isles*, in which 104 species of mosses and liveworts are mapped for the first time, was published this year.

D.W. Scott

References
Natural Environment Research Council (1978). *The generalised data management system G-EXEC*. London: NERC.
Smith, A.J.E. (1978). *Provisional atlas of the bryophytes of the British Isles*. Abbots Ripton: Biological Records Centre, Institute of Terrestrial Ecology.

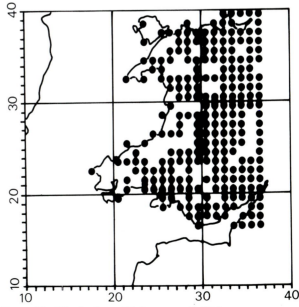

Figure 50 Distribution of Salix caprea *in Wales.*

REMOTE SENSING

For many years, ecologists have realised the value of aerial observation in the study of the nature and distribution of surface phenomena, and aerial observation in the form of photography taken from aircraft has proved a valuable tool in several studies undertaken by the Institute. The uses to which air photographs have been put in these studies have been very varied, and include :

1.   the provision of general overviews of a study area at the reconnaissance level :
2.   the provision of information on the main features of the surface phenomena in and around the study area, including their nature and spatial relationships ;
3.   the mapping of vegetation ;
4.   the counting of animals ; and
5.   the survey of coastal features.

Applications have up to now been focused on conventional aerial photography utilizing black and white, colour and colour infra-red (false colour) films, operating in the visible and near infra-red part of the electromagnetic spectrum. The potential for conventional air photography, utilizing the films mentioned above, for applications in terrestrial ecology is well-known and well-documented in ITE (e.g. Goodier 1971).

However, over the last decades, several new techniques and methods have been developed for the observation of surface phenomena, and are now available for use by the scientific community. These techniques and methods form part of the rapidly-expanding field of remote sensing. ITE has realised the necessity for attempting to keep up with these developments and the Data and Information Subdivision, in co-operation with the Geography Department of the University of Reading and other NERC Institutes, is exploring methods of improving the use of remote sensing within ITE. The initial phase of this co-operation has been the commissioning of a report on the current state of the remote sensing art and its application to the work of ITE. It is apparent from this report that, whilst new techniques and methods for environmental remote sensing have been developed in 2 areas namely :

1.   the sensing and recording of radiation from phenomena,
2.   the analysis and interpretation of remotely-sensed data,

the development in these 2 areas has not progressed in parallel. The rate of progress in the development of new sensors, platforms and methods of portrayal of remotely-sensed data has frequently outstripped the rate at which the ability to interpret the resulting data has grown. This disparity has led to remote sensing being criticised as a new-found technology looking for application, and it is true that the fundamental methodologies have yet to be established by which systems can be most profitably analysed and employed.

The major area for research within ITE may be in assessing the relevance of multivariate and other models to the analysis of digital or analogue information from remote sensing platforms. Already, work is in hand to develop suitable analytical models for investigating desertification in northern Kenya from LANDSAT data

C. Milner

Reference
Goodier, R. (1971). *The application of aerial photography to the work of the Nature Conservancy : proceedings of the Nature Conservancy staff seminar.* Edinburgh : The Nature Conservancy.

DATA BANKS

Many ecological data are not readily accessible, and ITE is attempting to remedy this difficulty in the form of computer data banks which can be readily searched and the information displayed. One example is the BRC data bank (see page 00) which is concerned with plant and animal distribution data. Two others are described here.

Site management information system
(This work was commissioned by the Nature Conservancy Council as part of its programme of research into nature conservation)

The Nature Conservancy Council has a considerable need for information to assist management at all levels. Studies have been undertaken, and are continuing, to assess this need and to try to define it in terms of systems, processing equipment and software requirements. At the regional level, one particular requirement is for information relating to sites, and this aspect is being considered in some detail for 2 purposes, namely, first, to develop, test and operate a system serving the needs of 1 or 2 regions as a means of formalising the requirement, and, second, to develop the experience of computer-based systems in a region.

Information requirements at 3 levels of site status are being considered : National Nature Reserves (NNR), sites graded 1 and 2 in the Nature Conservation Review, and sites of lesser status. For NNRs, the emphasis is upon servicing a project-based system for implementing the management plan. Proposals for individual projects, justified in terms of the management plan, are submitted by reserve management staff to the regional office. Arrangements for finance are made, subject to project approval, following a procedure for estimating 2 years in advance so that a series of project lists and ledgers is required to enable the management plan to operate sequentially from year to year. Close attention has been paid to accounting as a major activity within the regional office, and as a common link between almost all management projects. It is hoped that 2

```
NCR SITE BOUNDARY                    HABITATS                      OWNERSHIP
- - - - - - - - - - - - - - - - - - - - - - - - - - - - - - - - - - - - - - - - - -
I        53         54 I    I        53         54 I    I        53         54 I
I                                                                                 I
I            x          I    I            )          I    I            "          I
I          XXXXX        I    I          )))))        I    I          " " " "      I
I          XXXXX        I    I          )))))        I    I          " " "  "     I
I           XXXX        I    I           ))))        I    I           " "  "      I
I           XXX         I    I           )))         I    I           " "  "      I
I            X          I    I            )           I   I            "          I
I8          XX       8I   I8           ))        8I   I8          " "        8I
I9          XXX      9I   I9           )))       9I   I9          " " "      9I
I          XXXX         I    I          ))))         I    I          " "  "       I
I           XXX         I    I           )))         I    I           " "  "      I
I           XXX         I    I           )))         I    I           " "  "      I
I           XXX         I    I           )))         I    I           " "  "      I
I          XXXXX        I    I          )))))        I    I          " "  "  "    I
I         XXXXXX        I    I         ))))))        I    I         " "  " "  "    I
I        \XXXXXXXX      I    I        ))))))))))     I    I        " " "  "  "  "  I
I         XXXXXXXX      I    I         ))))))))      I    I         "  "  "  "  "  I
I7        XXXXXXX    7I   I7        ))))))     7I   I7        "  "  "  "   7I
I9        XXXX       9I   I9        ))))       9I   I9        "  " "  "    9I
I       XXXXX           I    I       )))))          I    I       "  " " "         I
I       XXXXX           I    I       )))))          I    I     ◁◁" " "            I
I        XXXX           I    I        ))))          I    I      ◁◁◁"              I
I        XXX            I    I        ))9           I    I      ◁◁◁                I
I         XX            I    I         )9           I    I       ◁◁                I
I         X          x  I    I         )          9 I    I       ◁         ◁      I
I         X        XXX  I    I         )        **9 I    I       ◁        ◁◁◁     I
I        XX        XXXX I    I        ))       ◆**9 I    I      ◁◁       ◁◁◁◁    I
I6        X        XXXX 6I   I6        )       ◆)*9 6I   I6     ◁        ◁◁◁◁  6I
I9       XX        XXXXX 9I  I9       ))      ◆◆)*9 9I  I9    ◁◁       ◁◁◁◁◁ 9I
I      XXXX XXXXXX      I    I      )))) )◆◆))*       I   I     ◁◁◁◁ ◁◁◁◁◁◁      I
I       XXXXXXXXX       I    I       )))))))))**      I   I      ◁◁◁◁◁◁◁◁◁      I
I         XXXXX         I    I         )))**          I    I         ◁◁◁◁◁        I
I         XXXXX         I    I        *))**           I    I         ◁◁◁◁◁        I
I         XXXXX         I    I        *****           I    I         ◁◁◁◁◁        I
I          XXX          I    I          ***           I    I          ◁◁◁         I
I          XX           I    I          )9            I    I          ◁◁          I
I          XXX          I    I          ))9           I    I          ◁◁◁         I
I5         XXX       5I   I5         ))9        5I   I5         ◁◁◁        5I
I9         XXX       9I   I9         ))9        9I   I9         ◁◁◁        9I
I          XX           I    I          ))            I    I          ◁◁          I
I          X            I    I          )             I    I          ◁           I
I          X            I    I          )             I    I          D           I
I          X            I    I          )             I    I          D           I
I          X            I    I          )             I    I          D           I
I          X            I    I          )             I    I          D           I
I          X            I    I          )             I    I          D           I
I          XX           I    I          ))            I    I          DD          I
I                                                                                 I
I        53         54 I    I        53         54 I    I        53         54 I
- - - - - - - - - - - - - - - - - - - - - - - - - - - - - - - - - - - - - - - - - -
```

) Mixed deciduous woodland — central facies   ◆ Beechwood
* Mixed deciduous woodland — lime facies   9 Neutral grassland

Figure 51 Symbolic representation of NCR habitats and ownership patterns in one hectare units of the Blackcliff-Syndcliff-Pierce Woods site. Details of individual habitats and owners are obtainable by reference to a directory file.
Habitat codes above are:
> *) mixed deciduous woodland—central facies*
> *◆ Beechwood *Mixed deciduous woodland—lime facies*
> *9—neutral grassland.*

important figures will emerge from operating in this way : i.e. the total cost of managing NNRs, identified by functions, and the minimum cost of keeping each reserve at its 'rock-bottom' level of management, which in effect means maintaining it within the legal constraints of its tenure. Further development will integrate the existing event record scheme, and will consider the feedback of biological records into the system.

For Grade 1 and 2 sites, attention has been concentrated on the requirements for planning at a regional/territorial level. Ownership and habitat maps for all the sites in Wales have been drawn up by the regions stored in the data banks and annotated with information relating to additional features, present status, opportunity for acquisition, suitable agency for acquisition, and threats. The data set derived from these maps will be used as a policy guide for site acquisition and as a core of information about the more important Welsh sites. This information can be displayed in a variety of ways, one of which is a map of the type illustrated in figure 51

The third level of site information relates to the degree of detail appropriate to SSSIs and other sites graded 3 and 4 in the Nature Conservation Review. An inventory approach is being considered, concentrating upon the major biological and physiographic attributes of a site and its conservation requirements. Many regional files relate to sites in this category and the need is for a simple retrieval system to help, for example, in the preparation of county schedules for SSSIs.

G.L. Radford

TERRESTRIAL ENVIRONMENT INFORMATION SYSTEM (TEIS)

This year has seen the completion of outline specifications for the Terrestrial Environment Information System (TEIS) which has been established at Bangor as the centre for terrestrial ecological data within NERC. Eventually, TEIS will constitute a single system which will link data derived from all ITE research and surveillance activities and which will provide connections to relevant data sources elsewhere. Initially, effort is being concentrated on the development of a referral capability, based on a detailed inventory of ITE data sources. Concurrently, an integrated data bank is to be compiled from suitable existing data sets as a pilot scheme. A programme is also being developed to survey and to specify in detail the requirements of users of ecological data in order to ensure that TEIS can satisfy these needs.
B.K. Wyatt

LIBRARY

The library service provides facilities and services in all 10 of the Institute's locations. The geographical dispersion of ITE creates problems and difficulties in

acquisitions policies, training and supervision of library personnel, and dissemination of information about the total resources of the library. Following a year in which efficient procedures were introduced for the acquisition and organisation of journals and books, attention has been turned to other problems. Acquisitions policies, for example, are extremely difficult to devise when many of the Institute's scientific interests are represented at most of its locations, and are dependent upon general acceptance that resources held in one library should be available throughout the Institute. Staff changes have focused on the need for establishing procedures during inter-regnums and for devising a training programme for new recruits. If only limited progress has been made on these topics, greater success can perhaps be claimed in the area of disseminating information about library resources. The monthly accessions bulletin of books received is also now being cumulated into an annual catalogue which will contain a subject index. It is hoped to use the PDP 11 computer at Merlewood to produce future bulletins and catalogues. A monthly "Current contents of books received" was produced to enhance the information provided in the accessions bulletins. A thesaurus of terms used in library catalogues, intended to aid both the indexing and retrieval of information contained in books, was completed.

A start of what promises to be a major development in library services within the Institute was made in the utilisation of on-line information retrieval services offered by the Department of Industry (DIALTECH) and Lockheed Information Systems (DIALOG). Preliminary assessment of the value of these services suggests that the main advantage to a library service with very limited staff-time for "manual" literature searches will be the speed of obtaining references relevant to the enquiry in hand.

J. Beckett

PUBLICATIONS

The Institute produced several publications in 1978. These include the first of a series of statistical checklists by the Director, entitled *Design of Experiments*. This leaflet is a checklist of 71 questions intended to ensure that a researcher uses experimental methods and analysis in the most efficient and unambiguous way. Although the leaflet is primarily intended for ecologists, it is relevant to other disciplines. In particular, it bridges the gap between a knowledge of statistical methods and their application to experimentation.

Another publication entitled *Overlays of environmental and other factors* provides a set of 12 overlays in a folder with explanatory text, for use with the Institute's Biological Records Centre distribution maps. Thus, the distribution of a species can be compared with factors such as climate, chalk, geology, altitude and various habitats. Each transparent overlay can be used independently with an overhead projector.

The booklet *Chemistry in ITE* describes the services offered by the Subdivision of Chemistry and Instrumentation, in support of the Institute's research. These services range from routine analysis and sampling to the development of sophisticated apparatus.

In addition, 6 provisional atlases of distribution maps have been produced covering the British Isles for mammals, bryophytes (mosses), Orthoptera (grasshoppers and crickets), Odonata (dragonflies), Trichoptera hydroptilidae (caddis flies), and Hymenoptera vespidae (social wasps). The final atlas for the bumblebees of the British Isles is expected to be published before this Report appears.

Other publications in press are *Virus diseases of trees and shrubs*, which describes virus diseases of native and indigenous trees and shrubs listed in alphabetical order of names; *The birds of St. Kilda*, recording 4 summers' fieldwork (1974–77) and summaries of previously published and unpublished records since 1959; *The distribution of fresh waters in Great Britain*, describing the number, distribution and size of lakes, rivers and reservoirs as part of a series of basic studies on the nature and extent of fresh waters in Great Britain; *An illustrated guide to river phytoplankton* in the same series as *A beginner's guide to freshwater algae* and *The Biological Records Centre* which describes the work of this unit. To date, 20 publications have been produced which have sold all over the world including the rest of Europe, North America, China, Africa and Australasia. In this way, the Institute hopes to provide a service for a wider readership than just the professional biologist and to fill gaps in the literature, especially in the educational field.

M.J. Woodman

HERBIVORE DENSITIES IN NORTHERN KENYA

This study has been partially funded by UNESCO/ UNEP under the auspices of the UNEP/MAB integrated project on arid land (IPAL). As indicated in last year's report, a project to develop mathematical models of the Hedad region of northern Kenya continues as part of the UNEP/MAB integrated project on arid lands. Initial models (figures 52-54) have now been constructed and provide a structured basis for further work.

Isohyets are available for the study area of Lake Turkana in northern Kenya and these data have been used to calculate the optimum density of herbivores in each of the 190 10 km squares into which the area has been divided.

Using Rosenzweig's (1968) predictive equation, with rainfall as the driving variable, it is possible to calculate primary production from

$$\log_{10}\text{NAAP} = \log_{10}\text{AE } (1 \cdot 66 \pm 0 \cdot 27) \\ - (1 \cdot 66 \times 0 \cdot 07)$$

where NAAP is the net above-ground annual primary production and AE is the annual actual evapotranspiration. In semi-arid and arid environments, it has been found that AE can be taken as being equal to the annual precipitation.

This model, which is similar to that of Phillipson (1975) and Coe *et al.* (1976), is inappropriate for dynamic simulation or optimisation since feedback between net primary production and herbivore densities (and hence vegetation offtake) is not considered. Nor are the spatial effects of animal movement interacting with vegetation or soil type considered in this initial state model. These feedback effects and some spatial influences have been incorporated into the second model which has some similarities to Goodall's model (1976),

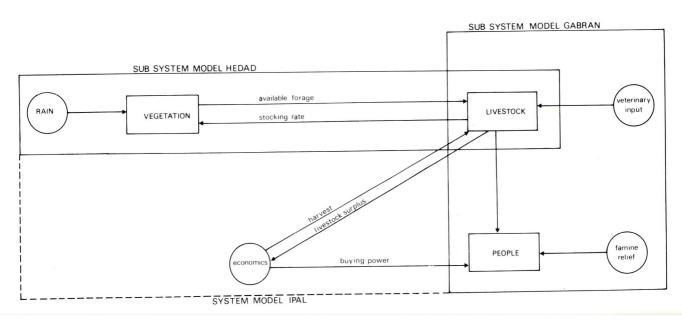

Figure 52 General inter-relationships within the Hedad pastoral system.

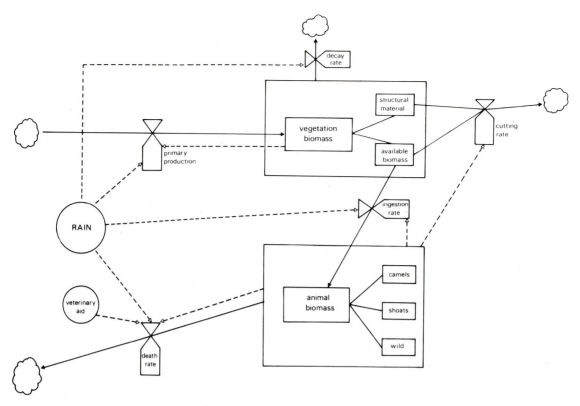

Figure 53   Submodel Hedad

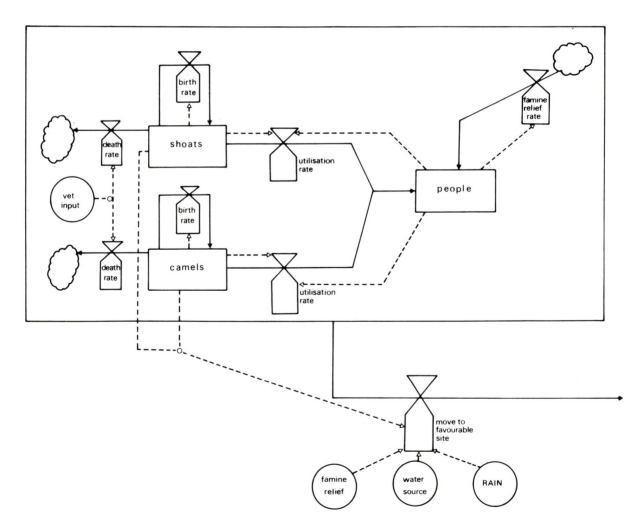

Figure 54.   Submodel Gabran

although there is a considerable scale difference. Initial experience with the new model has indicated the need for studies of some major processes within the desert ecosystem in the area investigated. The form of the animal/plant rate variable is the most important of these processes, although there is a need for some base-line state variable measurements such as the undisturbed mean biomass of major vegetation categories. The model is being continually updated in discussion with the main participants in the project, and, although some rate and state variable measurements are impossible to achieve with present resources, some changes in emphasis have been made.

In order to test the model against data collected in northern Kenya and also to make analysis rapid and accurate, a computer data bank has been set up for the project on the PDP 11/10 at Bangor and is a storage and retrieval system for small to medium data sets. The regular census data which are obtained from aerial survey of the study area are stored in the data bank and can be retrieved in either list or map form.

C. Milner

References
Coe, M.J., Cumming, D.H. & Phillipson, J. (1976). Biomass and production of large African herbivores in relation to rainfall and primary production. *Oecologia*, **22**, 341–354.
Goodall, D.W. (1976). Computer simulation of changes in vegetation subject to grazing. *J. Indian bot. Soc.*, **46**, 356–362.
Phillipson, J. (1975). Rainfall, primary production and "carrying capacity" of Tsavo National Park, Kenya. *E. Afr. Wildl. J.*, **13**, 171–201.
Rosenzweig, M.L. (1968). Net primary productivity of terrestrial communities: prediction from climatological data. *Amer. Nat.*, **102**, 67–74.

Chemistry and Instrumentation

Introduction
In the 5 years that the Institute has been established, the Subdivision has developed into an essential and valued component and its constituent units can all point to notable achievements, with little increase in total staff numbers. The success of the chemists can be measured by the doubling of analytical throughput, the engineers by the sophistication of the equipment constructed, the nursery staff by their ability to cope with the expansion in field and glasshouse experiments and, in addition, a small photographic service unit has been established.

During the past year, the most important single development in the Subdivision has been the radionuclide studies which are described in detail elsewhere in the report. Two other developments are also noteworthy. The first is the introduction of on-line data processing systems in the Subdivision's 2 chemical laboratories. The other is the big expansion in glasshouse facilities which is clearly going to have far-reaching implications for the staff of the Nursery Section.

Following the transfer of Mr. R. J. Parsell to NERC HQ the small laboratory devoted to electrophoretic analysis at Colney Research Station was closed.

THE SERVICE SECTIONS

Chemistry

1. Merlewood laboratory
    The Merlewood laboratory has again been heavily involved in the routine analyses arising out of the Institute's research projects. The principal requirement, as before, has been for the common nutrient and trace elements, and the laboratory is able to handle large numbers of samples. The increase in the numbers of water samples and pollution tests, mentioned in the 1977 report, has continued throughout the year, but there has been a marked drop in the number of soils analysed compared with 1977.

    The main feature of the year in relation to routine work has been the regular monthly intake of water samples, generated by the Scottish lochs survey, the Tayside synoptic survey and the regular monitoring of lakes and streams on nature reserves and sites of special interest to the Nature Conservancy Council. In the latter part of the year, samples of rainfall, canopy throughfall and tree stemflow were also examined. These were collected in Devilla Forest, Fife, in connection with the project concerned with atmospheric sulphur pollution. Because of storage and preservation problems, water samples must be analysed soon after receipt. A considerable part of each month has, therefore, had to be set aside for handling water samples and the routine soil and vegetation samples have had to be processed in short runs in the time available between the monthly water analyses. Careful planning and control have therefore been necessary to ensure that all samples were processed in the time allocated. In all, approximately 3000 water, 1500 soil and 2500 vegetation samples were examined during the year, involving over 60,000 individual tests. This work accounted for about 60 per cent of the time of the section staff.

    Estimations of fluorine are now routine, and about 2000 determinations were carried out during the year. The large majority of these determinations were on animal and bird material arising from the research on the effects of fluorine on predators. Some plant litter, vegetation and lichen samples were also processed.

    In addition to the water samples from Devilla Forest, other analyses were carried out for the sulphur pollution project. In this work, 600 filter paper discs, used to collect atmospheric particulate matter, were examined for total sulphur by X-ray fluorescence spectrometry.

The section acquired 2 new major pieces of equipment during 1978. The first was an organic carbon monitor for use in water analysis and the other was a dedicated on-line data processing system. Both these pieces of equipment are described later in the report.

J.A. Parkinson

2. Monks Wood laboratory

Following the report of the Visiting Group which came to Monks Wood in October 1977, an extra assistant was allowed for the analytical team and was appointed early in 1978. This staff increase, together with the new data processing facilities reported on later, have helped us to meet the continually-increasing demand for chemical analysis.

The Monks Wood laboratory concentrates on the analysis of organic pesticides and heavy metals in material originating largely from the Subdivision of Animal Function. The sample material is mainly of 2 types, animal (largely avian) tissue and water samples.

The emphasis in heavy metal analyses has been on lead, cadmium and mercury. Mercury is of particular interest because of its presence in organic forms. Techniques for both total and organic mercury are now in use. A much-needed improvement in precision in determining low concentrations of metals has been possible following the purchase of a new atomic absorption spectrophotometer background correction.

Work on organic pesticides, mainly organochlorines, has continued throughout the year, at a slightly lower level than in recent years. However, this reduction has been more than offset by the large numbers of samples analysed in connection with the research on monoterpenes. These compounds are examined at Monks Wood because of the skills in gas chromatography developed for organochlorine work.

M.C. French

Engineering

The Engineering Service includes a central unit at Bangor Research Station, where most of the complex construction jobs, especially in electronics, are carried out, and Station Engineers at Bush, Merlewood and Monks Wood. The year has been particularly difficult

for the engineering staff. The demands for construction and service work have consistently exceeded capacity to handle the work and some jobs have had to be refused, or delayed, with consequent frustrations all round. It is hoped that future project and staffing changes will ease the situation.

There has been a marked change in emphasis in 1978 from very large projects to smaller construction requirements. This contrasts with the preceding year when 2 of the jobs alone absorbed half the total engineering strength for the whole year. Examples of these smaller jobs include repairs to data logging equipment (Bangor), machining of parts for water flow experiments (Monks Wood), modifications to chemical instruments (Merlewood and Monks Wood) and repairs and modifications to growth cabinet and glasshouse control systems (Bangor and Bush).

The workshop accommodation has improved over the year. At Bush, a new workshop was completed and has been fitted with a comprehensive range of equipment. A new workshop has also been built at Brathens, but because there is no Station Engineer it is being used as a general facility. A general-use workshop has also been opened at Monks Wood, where staff can do simple jobs themselves under the supervision of the Station Engineer. The Bangor workshop, as at Bush, has obtained new equipment, the most notable being a storage oscilloscope and a universal bridge.

There have been no staff changes during the year, but the Bush and Monks Wood Engineers have received some assistance through temporary labour. At Bangor, a student, working under the industrial sandwich training scheme, successfully completed a number of small projects, including an automatic nickel-cadmium battery charger.

It is the practice of the Engineers to hold a group meeting at least once a year and in 1978 they all met at Monks Wood. Courses attended by the staff included one on micro-processor applications at Liverpool University (C. R. Rafarel) and one on workshop safety at Birmingham (G.H. Owen and V.W. Snapes). Two exhibitions were visited in the year, the annual Electronic and Instrumentation Exhibition (IEA) and an Engineering Design and Plant Maintenance Exhibition (PEMEC).

G.H. Owen

Nursery Unit

During the year, 3 new experimental areas were established at Bush.

The national birch collection site which involved fencing and sward establishment. The first part of the birch collection was planted in November.

2. The amenity grass mixture trial was established on an earlier trial site following ploughing and sterilising with methylbromide. This is an extension of previous work carried out in association with the Sports Turf Research Institute. The new trial was sown, established and carefully mown in stages down to a 2 cm sward within 2 months.

3. A standing ground-plunge bed area which had to be fenced and provided with irrigation facilities. This area was required to accommodate hundreds of potted plants needed in connection with studies to establish trees on industrial spoil heaps.

Bush tropical glasshouses have been equipped with warm (20°C) water storage tanks to overcome the problems of onset of dormancy and lowering of mist bed temperatures caused by cold water. There have been improvements in the clonal propagation of birch and alder cuttings in glasshouses using vernalised potted stock plants. Such stock, under extended daylight conditions, produces better quality cutting material sufficiently early in the summer to make a larger plant for over-wintering. The Senior Nurseryman has spent much of his time in the past year producing plans and specifications for new glasshouses in the Institute. Expanded glasshouse facilities are being developed at Merlewood, Bush, Banchory, Furzebrook and Monks Wood.

The landscaping plans for the SRC/NERC joint headquarters at Swindon were completed early in the year and are now being implemented. The site is heavily contaminated because of past industrial and railway use, with the result that more time and consideration have been needed than would normally be expected in a project of this sort.

The regular nursery jobs such as potting, watering and mowing inevitably take up much of the time of the staff of this section. However, new practices which might improve production are also tried. The use of herbicides is an interesting example. There have always been some reservations about using these substances in the field plots at Bush because of the risks of damage to experimental material, but 1 or 2 new products have been tried. In particular, glyptosate (Roundup) has been used successfully, and it has been possible to obtain positive control of *Agropyron repens* with no damage to woody material. This development is especially valuable for working amongst stock-plants and in closely planted areas.

R.F. Ottley

Photography

The small photographic unit based at Colney Research Station has continued to expand the range of services offered. In addition to a basic developing, printing and slide-copying service, the purchase of a new enlarger with filter head, 4-channel colour analyser and a print processing machine has enabled the unit to improve the colour printing service introduced last year. Colour prints from colour slides, using the Cibachrome colour reversal printing system, have so far formed the major part of the colour work undertaken.

The central photographic collection is now well-established and new material is being sought and added on a regular basis. An index system using small contact prints is being produced to improve access to the collection.

As well as expanding the services offered, the unit has been heavily involved in producing illustrations for reports, publications and displays, most notably for Station and conference exhibitions.

P.G. Ainsworth

RESEARCH AND DEVELOPMENT

Analytical evaluation

1. Total organic carbon analyser
   The extent of pollution in natural waters is frequently measured by well-tried procedures such as Chemical Oxygen Demand, Biological Oxygen Demand and Total Organic Matter. These are time-consuming techniques which have precipitated the development of instrumentation for the measurement of low levels of carbon in water. One such instrument for measuring both inorganic and organic carbon was obtained during 1978 and was modified and evaluated for operating in the range 0 to 10 mg l^{-1}. At 10 mg l^{-1} the precision was found to be satisfactory, with a relative standard deviation over 10 determinations of 2·5 per cent. The absolute detection limit was 0·05 mg l^{-1}. The response to different organic compounds at the same carbon level was found to vary, but the differences were not statistically significant. The equipment is now in routine use and is connected to an automatic sampler, allowing up to 50 samples to be processed daily.

2. On-line data processing
   To ease the pressure on the calculating facilities at the 2 chemical laboratories, automatic on-line data processing equipment has been installed at Merlewood and Monks Wood. The system in each case is based on a micro-processor with flexible disc storage, able to collect data from up to 8 analytical instruments simultaneously. At Merlewood, 5 lines are so far in use (2 automated colorimeters, 1 atomic absorption spectrophotometer and a dual channel flame instrument) and at Monks Wood the processor has been coupled to a gas chromatograph and atomic absorption spectro-

photometer. It is intended to extend the system to all instruments with suitable outputs in both laboratories.

The data are collected by a central unit and stored on a floppy disc. When an instrument run is finished, the sample and standard peak readings are computed and adjusted for base-line drift and standard correction.

In the Merlewood system, BASIC programmes can be used to further manipulate the data. These programming facilities are also available with limited disc storage to an operator at the same time as the collection of instrument data proceeds. Already this capacity is being used for the chemical data bank described elsewhere. A high quality printer is included in the system, and it is hoped to use the equipment at both Stations for final output of analytical reports.

3.   Boron procedures
Interest in the boron status of soils has resulted in an assessment of the analytical procedures available for the determination of this element. After investigation of a number of published methods, the well-known hot water extraction, followed by the methylene blue colorimetric procedure, was found to be the most suitable system. The procedure has been used for the assessment of reconstituted soils on industrial spoil, and has been found to be satisfactory over the range of $0 \cdot 1$ to 5 μg g^{-1}.

4.   Water sample preservation
An increasing demand for the analysis of water samples has highlighted the problems involved in handling procedures, in particular the delay time between collection and analysis. Previous studies on the effects of storage on chemical composition have been inconclusive and much of the literature on this subject is confusing. An investigation is, therefore, being carried out jointly by the Merlewood and Monks Wood chemical laboratories to examine a range of contrasting water types. Samples have been analysed immediately after collection, and aliquots were then treated with different preservatives and stored at different temperatures according to a factorial design. Further analyses were carried out after 21 and 90 days.

Preliminary results indicate that, in most cases, the major cations show very little change over 90 days. However, significant changes were noted for calcium and silicon after deep freezing. Labile constituents such as nitrate, phosphate and sulphate altered significantly over 21 days. No preservation method satisfactorily overcame this problem, but storage at 1°C slowed down the changes. Statistical evaluation of the data is continuing and the experiments are being extended to include rainwater.

5.   Muramic acid
Considerable interest is developing in the possibility of using muramic acid as an indicator of the amounts of micro-organisms in the soil. The Monks Wood laboratory is evaluating extraction and analytical techniques for determination by gas chromatography. The sample is first derivatised using Tri-sil Z in pyridine, and then injected into a gas chromatograph containing 3 per cent OV 17 on a 80–100 mesh Gas Chrom Q column. The chromatogram is run at 250°C and detection is achieved using flame ionisation. Results so far are encouraging, although problems still have to be solved at the soil extraction stage.

S.E. Allen, K.R. Bull, M.C. French, J.A. Parkinson and A.P. Rowland

Chemistry of aquatic pollutants

The chemical and physical behaviour of a pollutant in an aquatic ecosystem can greatly affect its availability to organisms. Consequently, the toxic effects of a pollutant on individuals or populations depend to a great extent on the chemical and physical properties of the pollutant and the nature of the aquatic environment itself. For this reason, chemical studies are being carried out in conjunction with those on fresh water organisms in the River Ecclesbourne, Derbyshire (see the Ecclesbourne project, pp 00) and in the continuous flow system at Monks Wood.

For the continuous flow experiments, a suitable water supply is essential. Such a supply must fulfil a number of criteria, not the least of which are its constant quality and its constant availability. Ground waters are often used for continuous flow work, but, in the absence of suitable geological formations beneath Monks Wood, alternatives had to be investigated. One alternative was to reconstitute purified water, and, as a result, a source of water for the experiments has been developed in which deionised water is remineralised in a carefully controlled fashion.

At an early stage in the development of the continuous flow system, a high capacity deionising plant was constructed at Monks Wood. This system has been used as the source of deionised water for the remineralisation plant. The plant, which operates automatically, provides a limited quantity (200 gallons day^{-1}) of experimental water in which the major cation and anion concentrations can be preselected and kept constant throughout the duration of an experiment. It was developed with 2 aims:

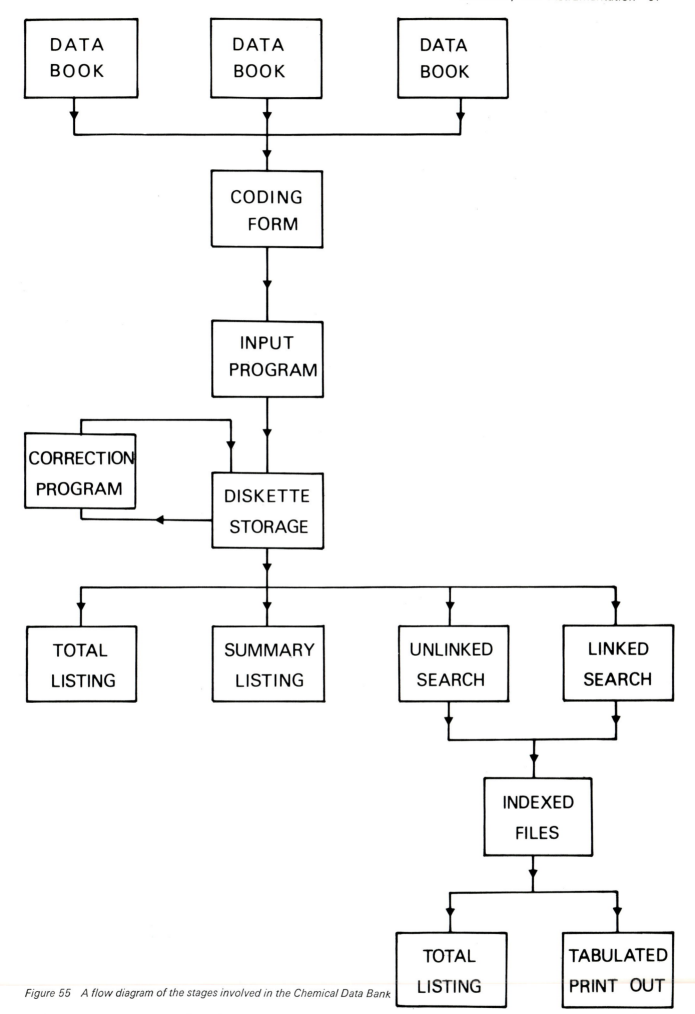

Figure 55   A flow diagram of the stages involved in the Chemical Data Bank

1. To provide an apparatus for supplying a wide range of waters, differing markedly in their composition, primarily for investigating the chemistry of aquatic pollutants in different waters; and

2. To provide a method for the large-scale production of remineralised water which would be suitable for extensive continuous flow experiments.

The major disadvantage with most remineralised waters, as often used for 'standard' waters in toxicity tests, is the limitation inherent in the use of soluble salts for the remineralisation process. Thus, for a hard water, solutions of calcium chloride and sodium bicarbonate are commonly mixed, which, although they give the desired concentrations of calcium and carbonate in solution, also give rise to generally undesirable concentrations of sodium and chloride ions. The mixing apparatus which has been developed is without this limitation. The method involves the dissolution of a calcium carbonate suspension with carbon dioxide gas and the subsequent removal of excess carbon dioxide with compressed air. Other cations and anions are added, as with other remineralisation methods, as soluble salts.

Some investigations were first carried out into the practicalities of running an automatic plant to produce remineralised water. It was decided to use a batch process in which a predetermined sequence of operations had to be carried out reliably and which could be altered as and when necessary. The process would also have to operate over a sufficiently small time-scale to provide water at an adequate rate. Finally, allowance had to be made for future experiments which required the presence of pollutants, including metals, in trace amounts. It therefore became essential to call upon the skills and knowledge of the Engineers to design and construct a suitable logic control system for the apparatus and to determine suitable materials and methods of construction for the mixing vessel itself. Some of this work has been mentioned in previous Institute reports.

The resulting plant has been installed at Monks Wood and has been supplying water for a variety of experimental purposes involving a number of different organisms. It has proved suitable for static experiments and for maintaining cultures of invertebrates and stocks of fish. However, although the plant is admirably suited to the first of its original aims, small-scale chemical studies, it has been decided that it is not practical to extend the plant for the large-scale production of water. Capital and running costs and maintenance demands upon manpower are prohibitive. For large-scale continuous flow work, it is planned to use nearby bore-hole facilities and to transport the water to the laboratories. The deionising plant at Monks Wood will enable dilution of this water to be made so as to achieve a suitable hardness for experiments. It is hoped that the bore-hole source will come 'on-line' in 1979.

K.R. Bull

Chemical data bank

Over the past few years, a chemical data bank has been available on the Merlewood PDP 8 computer. This system was reasonably comprehensive, but experience showed that a simpler version would be more useful. The data processing system already referred to has extensive diskette storage capacity and it was decided to adapt the data bank to this new data handling system.

The original system was designed for all chemical records including soils, rocks, waters, and plant and animal materials. Most interest has been shown in plant and animal material, and so initially only these data are being used, keeping an updated card index for rapid access to the other information. Data from the vegetation nutrient survey (see 1977 Annual Report) and from a programme of collection and analysis of less common species provide the basis for the vegetation data bank. Selected records will also be added from past, present and future analyses.

Keyboard input of records is directed to the central processing unit, which transfers data on to the diskette once the record is complete, and this record may later be checked by obtaining a print-out. A summary listing of batch number, date, site name and species name is available for filing purposes. Searching of the records may be performed on any characteristics (unlinked search) or combination of characteristics (linked search), with subsequent indexing of the relevant files. Output of these indexed files may be in the form of a total listing, or in tabulated form, including mean and standard deviation for each element. A flow diagram of the system is contained in figure 55.

A.P. Rowland and S.E. Allen

Chemical genetic indicators

Early studies in the Institute included the separation of flavonoids by thin-layer chromatography, and these were followed by provenance studies on *Pinus sylvestris* using monoterpenes separated by gas liquid chromatography. Variation in birch, and studies on population development in *Puccinellia* and *Agrostis* have been examined following the separation of isoenzymes. This separation was achieved using the well-established technique of electrophoresis in polyacrylamide.

The standard technique suffers from 2 disadvantages :—

1. Diffusion of the separated bands begins as soon as the electric field is switched off and continues as long as the enzyme activity is being assayed. The kinetics of diffusion of the assay reactants into and out of the gel is complex and makes quantitative assessments of doubtful value.

2. The assay generally has to terminate in a chromogenic reaction which is why oxido-reductase assays are so frequently used. Very many other

enzymes could be measured, but the necessary 2 or 3 stages of coupling to the nearest chromogenic reaction, though perfectly feasible in aqueous solution, become unreliable when carried out on the gel.

The present work aims to separate iso-enzymes by thin-layer, continuous flow iso-electric focusing, using beaded dextran as a mechanical support. This technique separates protein molecules in a pH gradient according to their iso-electric points and generally gives a 2- or 3-fold increase in the number of bands obtained by electrophoresis. By collecting the bands under the influence of the electric field, diffusion problems are reduced and the resulting aqueous fractions may be subjected to a greatly extended range of enzyme analysis. It is expected that kinetic studies could also be carried out. So far, the procedure has been tried with moderate success for medical studies, but there appears to have been little application for plant taxonomic purposes.

C. Quarmby

Engineering development

There was a decrease in development studies carried out in the Engineering laboratories in 1978 because of the reduced demand for major constructions compared with previous years. However, there was one new development area which will be of continuing interest. This is concerned with the application of micro-processors. The first piece of equipment produced using a micro-processor control system was a bird activity monitor for the Subdivision of Animal Function, in order to control the lighting duration in experimental rooms containing caged birds and to monitor bird activity at the same time. The equipment was designed and constructed to control the on-off times of the fluorescent lighting in each of the 10 experimental rooms. It was based on the use of thumb-wheel switches that could be preset at the required times. A M6800 micro-processor with VDU at Bangor was used to evaluate the system and the software was then developed. This was programmed into the control hardware so that it is now possible to scan the switch settings and to compare the information generated with the times shown on the external clocks. The lighting status for each room is then determined and the information used to operate the solid-state relays that control the lights. Bird activity in a particular cage is monitored by opto-electronic switches attached to perches in the cage. These events are summed over periods of time and then recorded on punched paper tape, together with lighting status, real time and room number.

Other tasks of special interest carried out by the Engineering staff during the year include the following:

1. Caving lamp dimmer—A method of conserving cap-lamp battery life, when searching for bats, was needed. This was achieved using pulse width modulation circuitry to control the supply to the lamp which can then be used at less than full brightness with considerable power economy.
2. Lake profile and transect recorder—Interfacing circuitry was developed to enable plots to be made of distances travelled by a boat in relation to water parameters.
3. Soil sieving machine—A cement mixer was modified by incorporating interchangeable screens. It enables bulk sieving of soils to be carried out with little effort.
4. Opto-electronic bat flight detecting window—The device records the movement in and out of a bat colony. An infra-red beam is emitted and reflected to and from 2 mirrors until picked up by a detector which records on a printing computer.
5. Boat modification—A portable boat has been modified to enable measuring transducers to be inserted through the base of the boat into the water. The instrument control structure has been mounted in the cabin.
6. Starling traps—Box traps were produced for trapping these birds, and included a novel door closure system based on the principles of a roller blind.
7. Metre quadrat—A new type of metre quadrat was devised allowing dimensions to be infinitely varied and which can be subdivided into 100 parts, and is much in demand.
8. Light radio meter—Developed to enable the light extinction coefficient to be measured at different underwater depths.

D.G. Benham, G.B. Elphinstone, G.H. Owen, C.R. Rafarel and V.W. Snapes

Birch germination

Advances in germination techniques have enabled the Nursery section to achieve rapid and more even germination of birch. Seedlings large enough to handle can now be obtained in 4 weeks at Bush. Although seed purity is 30 per cent, and germination varies from 5-40 per cent, seedling production is still rather unpredictable because of sensitivity to seed bed conditions, particularly moisture. The aim, therefore, was to provide stable conditions throughout the germination period with careful use of mist propagating facilities and supplementary lighting.

Five grams of the seed are sown on to a level sterilised potting compost contained in a 36×20 cm tray, and then covered with a light layer of 2-4 mm grit. No water is applied at this stage. The trays are then placed on the mist bench operating at a temperature range of

19–25°C which provides abundant moisture. 'Long-day' illumination is provided by a 400 watt lamp suspended 0·5 m above the grit surface. After about 10 days, germination is usually sufficiently well-advanced to allow the seedlings to be moved to a drier and cooler glasshouse with a temperature range of 15–20°C and no overhead mist. Illumination is continued under a different light intensity until the first true leaves have developed. This method is used between March and November, after which it is not possible to produce a plant large enough to be given an artificial 'winter' before the following spring.

R.F. Ottley and F.J. Harvey

Culture Centre of Algae and Protozoa

General review

The demand for cultures increased by 10 per cent over the year to a total of 4,073. This increase was probably in part a result of the passing of, or at least coming to terms with, the severe financial restrictions of recent years. An increase in the number and range of forms used for demonstration in universities may be some consolation to those who deplore the dearth of younger taxonomists. About 12 per cent of the output went to 36 countries overseas.

As usual, the number of new strains acquired has to be restricted, both on account of total work load, and the extra burden while new strains are unfamiliar. However, several new strains of *Tetrahymena* and *Euplotes* have been deposited by Dr. M. A. Gates of Toronto. These ciliate strains are particularly valued as they are playing an increasingly important role in many branches of biological research. Chief among the new strains are mating pairs of newly-described species which were formerly included in the aggregate species *Tetrahymena pyriformis*.

Basic taxonomic research into our organisms has continued to flourish. The main emphasis of protozoological research continues to be directed towards producing a guide to brackish and marine amoebae to follow the successful guide to fresh water and soil amoebae. Electron microscopical study on a species of amoeba has thrown new light on the relationship between different forms of the larger amoebae found in both fresh water and terrestrial habitats. Investigation into a widespread, but taxonomically neglected, group of slime moulds has revealed a possible relationship between them and the amoebae.

A comparative survey has just been completed on the fine structure of the flagellar base in chlorophyll C containing algae. It has considerable phylogenetic significance and will contribute to a better taxonomic understanding of what has long been a problem area to algal taxonomists.

Work on fresh water flagellates has included a description of a species of the commonly-seen genus *Bicosoeca* which is unique in accreting extraneous fragments of mainly organic material on the outside of an otherwise rather featureless lorica.

Recent observations of samples collected from various rivers mainly in the British Isles have revealed the existence of a considerable phytoplankton during the spring months composed largely of small centric diatoms.

Considerable interest has been shown in an account of a home-made miniature battery-operated centrifuge developed by Hilary Belcher. Very light and portable, it spins a pair of 1·5 ml polypropylene tubes at about 6,000 rpm and costs about £3 or less, according to what parts are already to hand.

Studies on marine and brackish flagellates in the collection have continued with the publication of 3 papers largely based on electron microscopical observation of the external features of *Pyramimonas*. It is now becoming clear that, with many flagellates, several species, and sometimes strains within species, have scales of characteristically-different patterns. This work will facilitate identification of those organisms which were otherwise rather featureless and often somewhat variable in shape.

Requests for information and advice continue to increase. Visitors have included several research students starting work with cultures and senior scientists from 12 different countries.

The outstanding progress in cryobiology has been reported on page 000.

E.A. George

Projects

listed by Subdivisions as at 8 February 1979

VERTEBRATE ECOLOGY SUBDIVISION   *code*

1   Monks Wood Experimental Station
2   Merlewood Research Station
3   Colney Research Station
4   Furzebrook Research Station
5   Edinburgh (Bush)
6   Edinburgh (Craighall Road)
7   Banchory (Brathens)
8   Banchory (Blackhall)
9   Bangor Research Station
10 Cambridge (Hills Road)
11 Cambridge (CCAP)
12 c/o UIV, Oxford

Project status
*   Supported by Nature Conservancy Council
†   Supported by Department of Environment
‡   Paid for by external contract
§   PhD Student supervised by ITE

54	Red deer ecology on Rhum	V.P.W. Lowe	2
59	Taxonomy of the red squirrel	V.P.W. Lowe	2
67	Prey selection in redshank	J.D. Goss-Custard	4
68	Dispersion in waders	J.D. Goss-Custard	4
104	Distribution and segregation of red deer	B.W. Staines	7
106	Red deer food studies	B.W. Staines	7
109	Annual cycles in Scottish red deer	B. Mitchell	7
111	Population dynamics of red deer at Glen Feshie	B. Mitchell	7
116	Freshwater survey of Shetland	P.S. Maitland	6
117*	Freshwater synoptic survey	P.S. Maitland	6
123	Zoobenthos at Loch Leven	P.S. Maitland	6
124	Fish distribution and conservation	P.S. Maitland	6
136	Hen harrier study in Orkney	N. Picozzi	8
138*	Puffin research	M.P. Harris	7
159	Upland bird project	D.C. Seel	9
208*	Amphibians and reptiles survey	H. Arnold	1
210*	Mammal distribution survey	H. Arnold	1
291*	Population ecology of bats	R.E. Stebbings	1
292*	Specialist advice on bats	R.E. Stebbings	1
322	Dispersal of otters	D. Jenkins	7
363	Dispersion of field voles in Scotland	W.N. Charles	6
385	Behaviour and dispersion of badgers	H. Kruuk	7
391	British mammals—the red fox	V.P.W. Lowe	2
420	Intraspecific variation in the Polar bear	V.P.W. Lowe	2
439	Red deer on the Isle of Scarba	B. Mitchell	7
441	Oystercatcher and shellfish interaction	J.D. Goss-Custard	4
442	Ecology of capercaillie	R. Moss	8
460	Interaction of gulls and puffins	M.P. Harris	7
461*	Puffins and pollutants	M.P. Harris	7
479	Red deer in production forests	B.W. Staines	7
498§	Wildcat studies	L.C. Corbett	7
499	Classification of Cervidae	V.P.W. Lowe	2
524	Fluoride in predatory mammals	K.C. Walton	8
525	Fluoride in predatory birds	D.C. Seel	8
528	Red deer in woodland habitats	B. Mitchell	7
543§	Population ecology of the red squirrel	V.P.W. Lowe	2
545*	Study of Lochs Morar and Shiel	P.S. Maitland	6
546‡	Study of Lochs Lomond, Awe and Ness	P.S. Maitland	6
555‡	Ecology of Craigroyston Woods	W.N. Charles	6
560	Red data book on vertebrates	H.R. Arnold	1
587*	Wild and feral deer in Great Britain	V.P.W. Lowe	2
588‡	Survey of seabirds in Victoria, Australia	M.P. Harris	7
606	Grey squirrel damage and management	R.E. Kenward	1
616‡	Baseline survey of Kilbirnie Loch	P.S. Maitland	6
617*	Synoptic survey of Yorkshire freshwaters	P.S. Maitland	6
627‡	Feral dogs and iguanas in Galapagos	H. Kruuk	7

INVERTEBRATE ECOLOGY SUBDIVISION   *code*

65	Invertebrate population studies	S. McGrorty	4
66	Variation in strandlines	S. McGrorty	4
161	Littoral fauna of Llyn Peris	A. Buse	9
162	Freshwater gastropods in North Wales	A. Buse	9
185	Effect of urbanisation	B.N.K. Davis	1

188	Woodland invertebrates	R.C. Welch	1
201	The white admiral butterfly	E. Pollard	1
202	The roman snail	E. Pollard	1
204*	Assessing butterfly abundance	E. Pollard	1
205	Invertebrates in hawthorn hedges	E. Pollard	1
211*	Lepidoptera distribution maps scheme	J. Heath	1
212*	Odonata distribution maps scheme	J. Heath	1
213*	Orthoptera distribution maps scheme	J. Heath	1
223	European invertebrate survey	J. Heath	1
229	Ecology/taxonomy of Spanish Hemiptera	M.G. Morris	4
230	Cutting experiment (Coleoptera)	M.G. Morris	4
231	Barton Hills grazing experiment (Coleoptera)	M.G. Morris	4
232	Butterfly studies at Porton Range	M.G. Morris	4
233	Cutting experiment (Hemiptera)	M.G. Morris	4
234	Grassland management by fire	M.G. Morris	4
236	Invertebrate populations in grass sward	E. Duffey	1
241	The fauna of box	L.K. Ward	1
243	Scrub succession at Aston Rowant NNR	L.K. Ward	1
255	Ecology of Myrmica species	G.W. Elmes	4
256	Protein electrophoresis	B. Pearson	4
261	Caste bias in Myrmica eggs	B. Pearson	4
282	Digestive enzymes	A. Abbott	4
270	Distributional studies on spiders	P. Merrett	4
273	Productivity of S. Magnus	N.R. Webb	4
274	Physiology of soil fauna	N.R. Webb	4
295	Survey of juniper in N. England	L.K. Ward	1
296	Scrub management at Castor Hanglands	L.K. Ward	1
309*	Phytophagous insects data bank	L.K. Ward	1
345	Spiders in East Anglian fens	E. Duffey	1
400	The large blue butterfly	J.A. Thomas	4
403	The black hairstreak butterfly	J.A. Thomas	4
404	The brown hairstreak butterfly	J.A. Thomas	4
405	Fauna of mature timber habitat	P.T. Harding	1
406	Non-marine Isopoda	P.T. Harding	1
407	British Staphylinidaes (Coleoptera)	R.C. Welch	1
414	Hartland Moor spider survey	P. Merrett	4
423§	Predator/prey relations on heathland	A.M. Nicholson	4
437‡	Further ecological studies on the Wash	S. McGrorty	4
450	Ecology of pseudo-scorpions	P.E. Jones	1
468*	Scottish invertebrate survey	E. Duffey	1
470	Upland invertebrates	A. Buse	9
474	Breckland open ground fauna	E. Duffey	1
500	Spiders on Hartland Moor NNR	P. Merrett	4
509§	Wood white butterfly population ecology	M. Warren	1
518	Myrmica sabuleti and M. scabrinodis	G.W. Elmes	4
527	Long-term changes in zooplankton	D.G. George	6
547	Study of the genus Micropteryx	J. Heath	1
557	Other distribution maps schemes	J. Heath	1
568	Subcortical fauna in oak	M.G. Yates	4
569	Insect fauna of Helianthemum & Genista	B.N.K. Davis	1
570	Studies on fritillary butterflies	E. Pollard	1
572	Aldabra management plan	M.G. Morris	4
577	Predation of freshwater zooplankton	D.H. Jones	6
592	Spatial organisation of zooplankton populations	D.G. George	6

ANIMAL FUNCTION SUBDIVISION   *code*

137*	Sparrowhawk ecology	I. Newton	6
178*	Causes of seabird incidents	I. Newton	6
179*	Heavy metals in waders	P. Ward	1

181*	Birds of prey and pollution	A.A. Bell	1
182	Aquatic herbicides	H.R.A. Scorgie	1
183	Frogs and pollution	A.S. Cooke	1
193	Stone curlew and lapwing	N.J. Westwood	1
199	Avian reproduction and pollutants	S. Dobson	1
289	Pollutants in freshwater organisms	F. Moriarty	1
325*	Carrion-feeding birds in Wales	I. Newton	6
413	Breeding biology of the cuckoo	I. Wyllie	1
435§	Social behaviour of thrushes	A. Tye	1
444	Endocrine lesions in birds	S. Dobson	1
455	Heavy metals in avian species	D. Osborn	1
455	Heavy metals and metabolism	D. Osborn	1
458	Shell formation and pollution	A.S. Cooke	1
473	Metal residues in birds of prey	A.A. Bell	1
475	Pollution and starling nutrition	P. Ward	1
559	Ecophysiology of the rabbit	D.T. Davies	1
590	Pollutants and the grey heron	J.W.H. Conroy	1

GROUSE AND MOORLAND ECOLOGY			code
129	Red grouse and ptarmigan populations	A. Watson	8
130	Management of grouse and moorlands	A. Watson	8
131	Golden plover populations	A. Watson	8
132	Monitoring in the Cairngorms	A. Watson	8
510§	Caecal threadworm and red grouse	G.R. Wilson	8

HEATHLAND SOCIAL INSECTS			code
252	Hartland Moor NNR survey	M.V. Brian	4
253	Tetramorium caespitum populations	M.V. Brian	4
258	Degree of control by queen ants	M.V. Brian	4
259	Larvae and worker communication	M.V. Brian	4
263	Worker ant activity	M.V. Brian	4
370	Inter-species competition in ants	M.V. Brian	4
371	Male production in Myrmica	M.V. Brian	4
578	Modelling an ant population	M.V. Brian	4

PLANT BIOLOGY SUBDIVISION			code
2	Meteorological factors in classification	E.J. White	5
73	Puccinellia maritima	A.J. Gray	4
81	Plant production, grazing and tree-line ecology	G.R. Miller	7
82	Seed produced by montane plants	G.R. Miller	7
102	Mountain vegetation populations	N.G. Bayfield	7
158	Community processes (physiology)	D.F. Perkins	9
160	Fluorine pollution studies	D.F. Perkins	9
208*	Botanical data bank	F.H. Perring	1
246	Physical environment, forest structure	E.D. Ford	5
247	Physiology of flowering	K.A. Longman	5
248‡	Physiology of root initiation	K.A. Longman	5
249	Morpho-physiological differences	M.G.R. Cannell	5
265	Regeneration on lowland heaths	S.B. Chapman	4
266	Root dynamics of Calluna vulgaris	S.B. Chapman	4
269	Autecology of Gentiana pneumonanthe	S.B. Chapman	4
329	Response of Scots pine	E.J. White	5
346	Genecology of grass species	A.J. Gray	4
359	Fibre yield of poplar coppice	M.G.R. Cannell	5
410	Tundra plants (bryophytes)	T.V. Callaghan	2
411	Taxonomy of bryophytes	S.W. Greene	5
412	Genecology of Racomitrium	B.G. Bell	5
451	Analysis of S. Georgian graminoids	T.V. Callaghan	2

506	Viruses of trees	J.I. Cooper	12
507	Ecologists' flora	E.M. Field	5

PLANT COMMUNITY ECOLOGY SUBDIVISION   *code*

1	Semi-natural woodland classification	R.G.H. Bunce	2
6	Scottish native pinewood survey	R.G.H. Bunce	2
9	Monitoring at Stonechest	J.M. Sykes	2
10	Monitoring at Kirkconnell Flow	J.M. Sykes	2
14	Tree girth changes in 5 NNR's	A.D. Horrill	2
48	Asulam effects on three upland pastures	A.D. Horrill	2
70	Management of sand dunes in East Anglia	L.A. Boorman	3
72	Salt marsh management	D.S. Ranwell	3
74	Sand dune stabilization	D.S. Ranwell	3
75	Control of Spartina	D.G. Hewett	9
77	Cliff vegetation methods	D.G. Hewett	9
78	Management of sand dunes in Wales	D.G. Hewett	9
92	Grazing intensities causing change	D. Welch	7
93	Assessing animal usage in N.E. Scotland	D. Welch	7
95	Importance of dung for botany change	D. Welch	7
120	Phytoplankton grazing & sedimentation	A.E. Bailey-Watts	6
121	Phytoplankton productivity	M.E. Bindloss	6
163	Ordination and classification methods	M.O. Hill	9
165	N. Wales bryophyte recording	M.O. Hill	9
225	Population studies on orchids	T.C.E. Wells	1
227	Sheep grazing on chalk grass flora	T.C.E. Wells	1
228	Effect of cutting on chalk grassland	T.C.E. Wells	1
242*	Establishment of herb-rich swards	T.C.E. Wells	1
318	Peat hydrology	A.J.P. Gore	1
340*	Survey of Scottish coasts	D.S. Ranwell	3
349	Maplin brent geese & wader studies	L.A. Boorman	3
360‡	Trees on industrial spoil	J.E. Good	9
362	Ecological survey of Cumbria	R.G.H. Bunce	2
364	Early growth of trees	A.H.F. Brown	2
367	The Gisburn experiment	A.H.F. Brown	2
369†	Sulphur content of tree leaves and bark	J.W. Kinnaird	7
374	Sand dune ecology in East Anglia	L.A. Boorman	3
377	Environmental perception studies	J. Sheail	1
380†	Monitoring of atmospheric SO2	I.A. Nicholson	7
381	Plankton populations in Loch Leven	D.G. George	6
388	Rusland Moss NNR survey	J.M. Sykes	2
389*	Management effect in lowland coppices	A.H.F. Brown	2
392†	Amenity grass cultivar trials	A.J.P. Gore	1
417	Silvicultural systems	A.H.F. Brown	2
424	Ecological survey of Britain	R.G.H. Bunce	2
452†	Foliar leaching and acid rain	J.W. Kinnaird	7
454*	NCC monitoring of woodlands	J.M.Sykes	2
463	Age class of amenity trees	J.E. Good	9
464	Popultn, competn and genetics of grasses	M.J. Liddle	1
466*	Ecology of railway land	J.M. Way	1
467	Roadside experiments	J.M. Way	1
483*	Scottish deciduous woodlands	R.G.H. Bunce	2
497*	Macrophyte studies	A.E. Bailey-Watts	6
549*	Monitoring in native pinewoods	J.M. Sykes	2
573‡	Amenity grass—stage 2	A.J.P. Gore	1
602	Modelling sports turf wear	T.W. Parr	1
605*	Phragmites dieback—phase III	L.A. Boorman	3

SOIL SCIENCE SUBDIVISION

code

4	Soil classification methods	P.J.A. Howard	2
8	Radiocarbon analysis of wood humus	A.F. Harrison	2
17	Meathop Wood IBP study	J.E. Satchell	2
21	Decomposition in Meathop Wood	O.W. Heal	2
22	Fungal decomposition of leaf litter	J.C. Frankland	2
29	Phosphorus circulation	A.F. Harrison	2
30	Biomass and decay of Mycena in Meathop Wood	J.C. Frankland	2
39	Phosphorus turnover in soils	A.F. Harrison	2
40	Woodland organic matter decomposition	P.J.A. Howard	2
44	Information handling and retrieval	D.K. Lindley	2
45	Tundra biome IBP study	O.W. Heal	2
61	Variation in growth of birch and sycamore	A.F. Harrison	2
87	Vegetation potential of upland sites	J. Miles	7
88	Plant establishment in shrubs	J. Miles	7
89	Calluna-Molinia-Trichophorum management	J. Miles	7
90	Birch on moorland soil and vegetation	J. Miles	7
118	Lake hydraulics	I.R. Smith	6
119	Physical limnology	I.R. Smith	6
140	Weathering and soil formation, Whin Sill	M. Hornung	9
148	Soil erosion on Farne Islands	M. Hornung	9
153	Mineralogical methods	A. Hstton	9
154	Field recording of profile data	M. Hornung	9
216	Register of NNRs	G.L. Radford	9
217	Species recording	G.L. Radford	9
218*	Event recording	G.L. Radford	9
219*	Data processing for BTO	D.W. Scott	1
221*	Data processing for Wildfowl Trust	D.W. Scott	1
245	Genetics of Betula nutrition	J. Pelham	5
302	Population growth and regulation	M.D. Mountford	10
306	Spatial processess and application	P. Rothery	10
307	Index of eggshell thickness	P.H. Cryer	10
308	Data from multi-compartment systems	P.H. Cryer	10
313	Seals research	M.D. Mountford	10
358	Earthworm production in organic waste	J.E. Satchell	2
384	Benthic microalgal populations	S.M. Coles	3
398	Upland land use	O.W. Heal	2
431	Soil change through afforestation	P.J.A. Howard	2
438	Ecology of Mycena galopus	J.C. Frankland	2
471	Soils of Upper Teesdale	M. Hornung	9
521	Mathematical modelling in Cumbria	D.W. Heal	2
522†	Ecology of vegetation change in uplands	D.F. Ball	9
533	Podzolic soils	P.A. Stevens	9
534	National land characterisation	D.F. Ball	9
541	Marginal land in Cumbria	C.B. Benefield	2
554	Cumbria land classes and soil types	J.K. Adamson	2
558	Fauna/mycoflora relationships	K. Newell	2
561	Soil fertility	M. Hornung	9
589	Microbial characteristics in soils	P.M. Latter	2
607*	Woodland soils conservation	D.F. Ball	9

DATA AND INFORMATION SUBDIVISION

code

365	Competition between grass species	H.E. Jones	2
376	Statistical training	C. Milner	9
402	Biometrics advice to NERC	M.D. Mountford	10
421	Management information system development	D.I. Thomas	1
432	Effect of birch litter on earthworms	J.E. Satchell	2
434	ITE computing services	C. Milner	9

457	Grazing models	C. Milner	9
494	Computing facilities at Hope Terrace	I.R. Smith	6
496	Data processing services at Monks Wood	D.W. Scott	1
512	National collection of birch	A.S. Gardiner	2
514	British birch publication	A.S. Gardiner	2
528	Biological data bank	D.W. Scott	1
530	Laser scan mapping system	D.W. Scott	1
531	Statistical & computing advice, Furzebrook	R.T. Clarke	4
548	Leaf-shape analysis of European birch	A.S. Gardiner	2
556	Estimation in acid rain	K.H. Lakhani	1
564	British Hydracarina—mainly of mosses	N. Hamilton	2
565	Bibliography of Shetland	N. Hamilton	2
566	Islands biogeographic analysis	N. Hamilton	2
574	Potential for fuel cropping in upland Wales	D.I. Thomas	9
579	Woodland research conference	A.S. Gardiner	2
591	Terrestrial Environment Information System	B.K. Wyatt	9
598	Information retrieval system for Dorset heaths	R.T. Clarke	4
603	Measures of familial similarity	P. Rothery	10
604[†]	Development of VIEWDATA facility	B.K. Wyatt	9
609	Biological classn of rivers in the UK	D. Moss	9
621	Models of rabies epidemiology	P.J. Bacon	2

CHEMISTRY AND INSTRUMENTATION SUBDIVISION *code*

23	Soil temperature in Meathop Wood	K.L. Bocock	2
52	Biological studies of Glomeris	K.L. Bocock	2
62	National plant nutrient survey	H.M. Grimshaw	2
378	Chemical data bank	S.E. Allen	2
481	Monitoring pollutants in natural waters	K. Bull	1
482	Chemistry of aquatic pollutants	K. Bull	1
484	Chemical technique development	Parkinson/French	2
485	Chemical support studies	S.E. Allen	2
486	Engineering development	G.H. Owen	9
487	Microprocessor development studies	C.R. Rafarel	9
489	Glasshouse and nursery maintenance	R.F. Ottley	5
490	Photographic development	P.G. Ainsworth	3
491	Isotope development studies	S.E. Allen	2
553[†]	Radionuclide contamination of ecosystems	K.L. Bocock	2

CULTURE CENTRE OF ALGAE AND PROTOZOA *code*

445	Marine flagellates taxonomy	J.H. Belcher	11
446	Freshwater flagellates taxonomy	D.J. Hibberd	11
447	Freshwater and marine amoebae	F.C. Page	11
448	Colourless flagellates taxonomy	E.M.F. Swale	11
449	Preservation of cultures	G.J. Morris	11

DIRECTORATE *code*

203	The cinnabar moth	J.P. Dempster	1
393	The swallowtail butterfly	J.P. Dempster	1
408[†]	Arboriculture: selection	F.T. Last	5
503	Development of systems analysis	J.N.R. Jeffers	2
504	Markov models	J.N.R. Jeffers	2
505	Ecology of Outer Hebrides	J.N.R. Jeffers	2
508	Botanical variation in elm	J.N.R. Jeffers	2
511	Landscaping at Swindon	F.T. Last	5
516	Forest management for energy	R.C. Steele	10
517	Primary productivity in woodlands	J.N.R. Jeffers	2
518[‡]	UNESCO MAB information system	J.N.R. Jeffers	2
526[‡]	Monitoring in Banff and Buchan	F.T. Last	5

Staff List March 1979

Institute of Terrestrial Ecology
Address as for
Merlewood Research Station

Director
Mr J.N.R. Jeffers   CSO

Institute of Terrestrial Ecology
68 Hills Road
Cambridge
CB2 1LA
0223 (Cambridge) 69745-9
Telex 817201

Senior Officer
Head of Division of Scientific Services
DCSO   Mr Steele, R.C.
PS   Mrs Waterfall, G.M.

Institute Secretary
Prin   Mr Ferguson, J.G.

Administration,Finance and Establishments
SEO   Mr Collins, R.T.
HEO   Mr Clapp, E.C.J.
EO   Miss Boyden, B.R.
EO   Mrs Chrusciak, W.
CO   Miss Barrett, L.M.
CO   Mrs Cooke, I.P.
CO   Miss Hunt, R.J.
CO   Mr Taylor, A.C.E.
Aud/t   Mrs Chambers, E.M. (PT)

DIVISION OF SCIENTIFIC SERVICES
Subdivision of Data and Information
PSO   Mr Mountford, M.D.
SSO   Mr Cryer, P.H.
SSO   Mr Rothery, P.
HSO   Mr Spalding, D.F.

Publications and Liaison Officer
PSO   Mr Woodman, M.J.

Institute of Terrestrial Ecology
c/o Unit of Invertebrate Virology
5 South Parks Road
Oxford
OX1 3UB
0865 (Oxford) 52081

DIVISION OF PLANT ECOLOGY
Subdivision of Plant Biology
SSO   Dr Cooper, J.I.
SO   Dr Edwards, Mary L. (PT)
ASO   Mrs McCall, D. (PT)

Institute of Terrestrial Ecology
Monks Wood Experimental Station
Abbots Ripton
Huntingdon
PE17 2LS
048 73 (Abbots Ripton) 381-8
Telex 32416

Senior Officer
Head of Division of Animal Ecology
DCSO   Dr Dempster, J.P.

PS   Mrs Stocker, B.J.

Administration
HEO   Mr Cheesman, J.A.
CO   Mrs Burton, V.J.
CO   Mr Cotton, A.E.
CO   Mrs Grihault, S.M.
CO   Mrs Haas, M.B. (PT)
CO   Mrs Wood, H.
CA   Miss Hall, J.R.
Typ   Mrs Plant, D.S.
Aud/t   Mrs Glover, P.R.
Aud/t   Mrs Stokes, J.
Clnr   Mrs Bell, K.C.J. (PT)
Clnr   Mrs Chance, M.E. (PT)
Clnr   Mrs Ennis, S. (PT)
Clnr   Mrs McDowell, J. (PT)
Clnr   Mrs Schietzel, P.E. (PT)
Band 8   Mr Farrington, T.F.
Band 6   Mr Baker, A.W.
H/kpr   Mrs West, M.K.

DIVISION OF ANIMAL ECOLOGY
Subdivision of Vertebrate Ecology
SSO   Dr Stebbings, R.E.
HSO   Dr Kenward, R.E.
SO   Mr Arnold, H.R.

Subdivision of Invertebrate Ecology
PSO   Dr Davis, B.N.K.
PSO   Dr Duffey, E.A.G.
PSO   Mr Heath, J.
PSO   Dr Pollard, E.
PSO   Dr Ward, Lena K.
PSO   Dr Welch, R.C.
HSO   Mr Harding, P.T.
HSO   Mrs Welch, J.M.
SO   Mr Jones, P.E.
SO   Mrs King, M.L.
SO   Mr Moller, G.J.
ASO   Miss Brundle, H.A.
ASO   Mr Greatorex-Davies, J.N.
ASO   Mr Plant, R.

Subdivision of Animal Function
SPSO   Dr Newton, I. *Head of Subdivision*
                       (located in Edinburgh until
                       August 1979)

PSO   Dr Moriarty, F.
SSO   Dr Dobson, S.
SSO   Mr Westwood, N.J.
HSO   Mr Bell, A.A.
HSO   Mr Conroy, J.W.H.
HSO   Dr Davies, D.T.

HSO   Mr Dawson, A.S.
HSO   Dr Osborn, D.
HSO   Dr Scorgie, H.R.A.
SO   Ms Hanson, H.M.
SO   Mr Wyllie, I.
ASO   Miss Brown, M.C.
ASO   Mr Howe, P.D.
ASO   Miss Ward, J.R.
ASO   Mr Myhill, D.
Band 4   Mr Thomson, H. (PT)
Band 4   Mrs Wade, E.E.C. (PT)

Subdivision of Plant Community Ecology
SPSO   Mr Gore, A.J.P. *Head of Sub-
                          division*

PSO   Dr Hooper, M.D.
PSO   Dr Sheail, J.
PSO   Dr Way, J.M. (on secondment to
                       DOE)

PSO   Mr Wells, T.C.E.
HSO   Dr Sargent, Caroline M.
HSO   Mr Frost, A.J.
HSO   Mr Parr, T.W.
SO   Miss Cox, R.
SO   Mr Lowday, J.E.
ASO   Mrs Bell, S.A.
ASO   Mr Mountford, J.O.

DIVISION OF SCIENTIFIC SERVICES
Subdivision of Data and Information
PSO   Mr Lakhani, K.H.
SSO   Mrs Greene, D.M.
SSO   Miss Scott, D.W.
SDP   Miss Dodson, S.D.
DP   Mrs Binge, C.

Library
ALIB   Mrs King, K.B. *Deputy Librarian*
CO   Mrs Purdy, M.I.

*Subdivision of Chemistry and
Instrumentation*
SSO   Mr French, M.C.
SSO   Dr Bull, K.R.
SO   Mr Freestone, P.
ASO   Mr Leach, D.V.
ASO   Mr Sheppard, L.A.
P&TO3   Mr Snapes, V.W. (Workshop)

Institute of Terrestrial Ecology
Merlewood Research Station
Grange-over-Sands
Cumbria
LA11 6JU
044 84 (Grange-o-Sands) 2264-6
Telex 65102

Director, ITE
CSO   Mr Jeffers, J.N.R.
EO   Mrs Ward, P.A.
Typ   Mrs Woodward, D.T.

Senior Officer
Head of Subdivision of Soil Science
SPSO   Dr Heal, O.W.
PS   Miss Duncan, H.R.

Administration
HEO   Mrs Foster, E.
CO   Mrs Coward, P.M.
CO   Miss Hunt, J. (PT)
CA   Miss Legat, C.R. (PT)
Sh/t   Miss Benson, V.E.
Typ   Mrs Kay, C.G. (PT)
Typ   Mrs Sigrist, W.A.
Typ   Miss Stewart, S.
Clnr   Mrs Burton, E. (PT)
Clnr   Mr Casey, M. (PT)
Clnr   Mrs Pearson, V. (PT)
Band 8 Mr Foster, P.L.
Band 4 Mr Gaskarth, J.

DIVISION OF ANIMAL ECOLOGY
Subdivision of Vertebrate Ecology
PSO   Mr Lowe, V.P.W.

DIVISION OF PLANT ECOLOGY
Subdivision of Plant Biology
SSO   Dr Callaghan, T.V.
HSO   Mr Lawson, G (C/T)
HSO   Mr Scott, R
ASO   Miss Whittaker, H.A. (C/T)

Subdivision of Plant Community Ecology
PSO   Mr Brown, A.H.F.
PSO   Dr Bunce, R.G.H.
PSO   Mr Sykes, J.M.
SSO   Dr Horrill, A.D.
SSO   Mr Millar, A.
SO   Mr Barr, C.J.
SO   Miss Robertson, S.M.C.
ASO   Mr Briggs, D.R.
ASO   Miss Conroy, C.L. (C/T)
ASO   Miss Dickson, K.E. (PT)

Subdivision of Soil Science
SPSO   Dr Heal, O.W. *Head of Subdivision*

PSO   Mr Howard, P.J.A.
PSO   Dr Satchell, J.E.
SSO   Dr Frankland, Juliet C. (C/T PT)
SSO   Dr Harrison, A.F.
HSO   Mr Bailey, A.D.
HSO   Mr Benefield, C.B.
HSO   Dr Dighton, J.
HSO   Miss Latter, P.M.
SO   Mr Adamson, J.K.
SO   Mrs Howard, D.M.
SO   Mrs Howson, G. (PT)
SO   Mrs Shaw, F.J.
SO   Mr Smith, M.R.
ASO   Miss Costeloe, P.L. (PT)
ASO   Mr Nelson, A.
ASO   Miss Martin, K.J.

DIVISION OF SCIENTIFIC SERVICES
Subdivision of Data and Information
PSO   Mr Lindley, D.K.
SSO   Mr Gardiner, A.S.
SSO   Dr Jones, Helen E. (C/T PT)
HSO   Mr Bacon, P.J.
SO   Mrs Adamson, J.M.
SO   Miss Hamilton, N.M.

Library
LIB   Mr Beckett, J. *Chief Librarian*
CO   Mrs Killalea, M.A.

Subdivision of Chemistry and Instrumentation
PSO   Mr Allen, S.E. *Head of Subdivision*

PSO   Mr Bocock, K.L.
SSO   Mr Grimshaw, H.M.
SSO   Mr Parkinson, J.A.
SSO   Mr Quarmby, C.
HSO   Mr Benham, D.G.
HSO   Mr Roberts, J.D.
SO   Mr Rowland, A.P.
ASO   Mrs Benham, P.E.M.
ASO   Mr Coward, P.A.
ASO   Mrs Kennedy, V.H. (PT)
ASO   Mrs Whittaker, M.
ASO   Mrs Rigg, J.
ASO   Mrs Zirkel, S.M. (P/T)

Institute of Terrestrial Ecology
Furzebrook Research Station
Wareham
Dorset
BH20 5AS
092 93 (Corfe Castle) 361-2
Telex 418326

Senior Officer
Head of Subdivision of Invertebrate Ecology
SPSO   Dr Morris, M.G.

Administration
EO   Mr Currey, R.J.
CO   Mrs Perkins, M.K.
CO   Miss Agate, E.J.
Typ   Miss Richmond, W.
Clnr   Mrs Fooks, N.M. (PT)
Clnr   Mrs Jeans V.V. (PT)
Band 6 Mr Jeans A.G.

DIVISION OF ANIMAL ECOLOGY
Subdivision of Vertebrate Ecology
PSO   Dr Goss-Custard, J.D.
ASO   Miss Durell, S.E.A.

Subdivision of Invertebrate Ecology
SPSO   Dr Morris, M.G. *Head of Division*

PSO   Dr Merrett, P.
PSO   Dr Webb, N.R.C.
SSO   Dr Elmes, G.W.
SSO   Dr McGrorty, S.
SSO   Dr Thomas, J.A.
HSO   Mr Abbott, A.M.
HSO   Mr Pearson, B.
HSO   Mr Snazell, R.G.
HSO   Mr Reading, C.J.
SO   Mr Rispin, W.E.
SO   Mr Yates, M.G.
ASO   Mrs Jones, R.M.
ASO   Mrs Wardlaw, J.C.

Special Merit: Heathland Social Insects
SPSO   Dr Brian, M.V.

DIVISION OF PLANT ECOLOGY
Subdivision of Plant Biology
PSO   Dr Chapman, S.B.
PSO   Dr Gray, A.J.
SSO   Dr Daniels, R.E.
ASO   Mr Rose, R.L.

DIVISION OF SCIENTIFIC SERVICES
Subdivision of Data and Information
SO   Mr Clarke, R.T.

Institute of Terrestrial Ecology
Colney Research Station
Colney Lane, Colney
Norwich, Norfolk
NR4 7UD
0603 (Norwich) 54923-5

Senior Officer
PSO   Dr Ranwell, D.S.

DIVISION OF PLANT ECOLOGY
Subdivision of Plant Community Ecology
PSO   Dr Boorman, L.A.
SO   Mr Fuller, R.M.
ASO   Mr Storeton-West, R.L.

DIVISION OF SCIENTIFIC SERVICES
Subdivision of Chemistry and Instrumentation
SO   Mr Ainsworth, P.G.

Institute of Terrestrial Ecology
Bush Estate
Penicuik
Midlothian
EH26 0QB
031 445 4343-6
Telex 72579

Senior Officer
Head of Division of Plant Ecology
DCSO   Prof Last, F.T.
PS   Mrs Hogg, A.H.

Administration
HEO   Mr Lally, P.B.
CO   Miss Maxwell, M.
CA   Mrs Campbell, A.M. (PT)
Typ   Miss Thomson, L.S.
Clnr   Mrs Mowat, E.A.M.
Clnr   Mrs Innes, D.S. (PT)

DIVISION OF PLANT ECOLOGY
Subdivision of Plant Biology
SPSO   Dr Greene, S.W. *Head of Subdivision*
PS   Mrs McHugh, E.

PSO   Dr Cannell, M.G.R.
PSO   Dr Ford, E.D.
PSO   Dr Longman, K.A.
SSO   Mr Deans, J.D.
SSO   Dr Fowler, D.

SSO	Dr Leakey, R.R.B.
SSO	Dr Milne, R.
SSO	Mr White, E.J.
HSO	Mr Bell, B.G.
HSO	Ms Field, E.M.
SO	Miss Cahalan, C.M.
SO	Mrs Lamont, L.C.
SO	Mr Lightowlers, P.J.
SO	Mrs Macleod, J.
SO	Mr Murray, T.D.
SO	Miss Wilson, J. (C/T)
ASO	Mr Davies, S.J.
ASO	Miss Dick, J. McP.
ASO	Mrs Halcrow, A.
ASO	Mr Leith, I.D.
ASO	Mr Wilson, R.H.F.
ASO	Mrs Wilson, N. (C/T)

Subdivision of Plant Community Ecology
ASO   Mr Munro, R.C.

Subdivision of Soil Science
PSO   Mr Pelham, J.
SSO   Dr Mason, P.A.
ASO   Mrs Fowler, A.F.O.
ASO   Mr Ingleby, K.

DIVISION OF SCIENTIFIC SERVICES
Subdivision of Data and Information
HSO   Mr Smith, R.I.

Library
HSO   Mr Melville-Mason, G.N.L.I.
CO   Mrs Shields, S.E. (PT)

Subdivision of Chemistry and Instrumentation
HSO   Mr Ottley, R.F. *Snr Nurseryman*

SO   Mr Harvey, F.J.
P&TO4   Mr McCormack, J.W.
P&TO4   Mr Elphinstone, G.B.

Institute of Terrestrial Ecology
78 Craighall Road
Edinburgh
EH6 4RQ
031 552 5596

Administration
CO   Mrs Adair, S.M. (PT)
Typ   Mrs Wilson, S.M.

DIVISION OF ANIMAL ECOLOGY
Subdivision of Animal Function
Head of Subdivision
SPSO   Dr Newton, I. (transfers to Monks Wood August 1979)

Subdivision of Vertebrate Ecology
PSO   Dr Maitland, P.S.
SSO   Mr Charles, W.N.
SSO   Mr East, K.
HSO   Dr Marquiss, M.
SO   Mrs Duncan, P. (C/T)
SO   Mr Morris, K.H.
SO   Ms Smith, B.D. (C/T)
ASO   Mr Rosie, A.J. (C/T)

Subdivision of Invertebrate Ecology
SSO   Mr Jones, D.H.

DIVISION OF PLANT ECOLOGY
Subdivision of Plant Community Ecology
SSO   Dr Bailey-Watts, A.E.
SSO   Dr Bindloss, Margaret E.
ASO   Mr Kirika, A.

DIVISION OF SCIENTIFIC SERVICES
Subdivision of Data and Information
PSO   Mr Smith, I.R.
ASO   Mr Lyle, A.A.

Institute of Terrestrial Ecology
Hill of Brathens
Glassel
Banchory, Kincardineshire
AB3 4BY
033 02 (Banchory) 3434
Telex 739396

Senior Officer
Head of Subdivision of Vertebrate Ecology
SPSO   Dr Jenkins, D.

Administration
EO   Miss Reffin, S.J.
CO   Miss Pirie, A.
CO   Mrs Stevenson, M.P.
Typ   Miss Anderson, V.E.
Clnr   Mrs Griffin, M.D. (PT)
Clnr   Mrs Ritchie, R. (PT)
Band 8   Mr Griffin, C.

DIVISION OF ANIMAL ECOLOGY
Subdivision of Vertebrate Ecology
SPSO   Dr Jenkins, D. *Head of Subdivision*

PSO   Dr Harris, M.P.
PSO   Dr Kruuk, H.
PSO   Dr Mitchell, B.
PSO   Dr Staines, B.W.
SSO   Mr McCowan, D.
HSO   Mr Parish, T.
SO   Mr Catt, D.C.
SO   Miss Harper, R.J.

DIVISION OF PLANT ECOLOGY
Subdivision of Plant Biology
PSO   Dr Miller, G.R.
SSO   Dr Bayfield, N.G.
HSO   Mr Cummins, R.P.

Subdivision of Plant Community Ecology
PSO   Mr Nicholson, I.A.
SSO   Mr Kinnaird, J.W.
SSO   Mr Welch, D.
HSO   Mr Paterson, I.S.
ASO   Mrs Cummins, C.M.
ASO   Miss McPherson, M. (C/T)

Subdivision of Soil Science
PSO   Dr Miles, J.
SO   Mr Young, W.F.

DIVISION OF SCIENTIFIC SERVICES
Subdivision of Data and Information
HSO   Mr French, D.D.

Institute of Terrestrial Ecology
Blackhall
Banchory
Kincardineshire
AB3 3PS
033 02 (Banchory) 2206-7

Administration
Typ   Mrs Allan, E.J.P.

DIVISION OF ANIMAL ECOLOGY
Special Merit: Grouse and Moorland Ecology
SPSO   Dr Watson, A.

Subdivision of Vertebrate Ecology
PSO   Dr Moss, R.
SSO   Mr Picozzi, N.
HSO   Mr Parr, R.A.
ASO   Mr Glennie, W.W.
ASO   Mr Watt, D.C.

Institute of Terrestrial Ecology
Bangor Research Station
Penrhos Road
Bangor, Gwynedd
LL57 2LQ
0248 (Bangor 4001-5
Telex 61224

Senior Officer
Head of Subdivision of Data and Information
SPSO   Dr Milner, C.
PS   Mrs Lloyd, A.C.

Administration
HEO   Mr Jones, W.L.
CO   Miss Evans, W.
CO   Mrs Thomson, J.A.
CA   Miss Owen, D.E.
Typ   Miss Roberts, M.E.
Typ   Mrs Bennison, M.I.
Clnr   Mrs Jones, O.M. (PT)
Clnr   Mrs Stedmond, L.A.
Band 4   Mr Wilson, J.N.

DIVISION OF ANIMAL ECOLOGY
Subdivision of Vertebrate Ecology
SSO   Dr Seel, D.C.
SSO   Mr Walton, K.C.
HSO   Mr Thomson, A.G.

Subdivision of Invertebrate Ecology
SSO   Dr Buse, A.

DIVISION OF PLANT ECOLOGY
Subdivision of Plant Biology
PSO   Dr Perkins, D.F.
SSO   Mrs Jones, V.
HSO   Mr Millar, R.O.
SO   Mrs Neep, P.

Subdivision of Plant Community Ecology
PSO   Mr Hill, M.O.
PSO   Mr Shaw, M.W.
SSO   Mr Dale, J.
SSO   Dr Good, J.E.G.

SSO	Mr Hewett, D.G.
HSO	Mr Evans, D.F.
HSO	Miss Pizzey, J.M.
ASO	Miss Bellis, J.A. (C/T)
ASO	Mrs Hays, J.A.

Subdivision of Soil Science

PSO	Dr Ball, D.F.
PSO	Dr Hornung, M.
HSO	Mr Williams, W.M.
SO	Miss Hatton, A.A.
SO	Mr Stevens, P.A.
ASO	Mr Hughes, S.

DIVISION OF SCIENTIFIC SERVICES
Subdivision of Data and Information

| SPSO | Dr Milner, C. *Head of Subdivision* |

SSO	Mr Radford, G.L.
SSO	Mr Thomas, D.I.
SSO	Dr Wyatt, B.K.
HSO	Dr Moss, D.
ASO	Miss Grady, W.
DP	Miss Jones, M.L.

Library

| CO | Mrs Owen, M. |

Subdivision of Chemistry and Instrumentation

| SSO | Mr Owen, G.H. *Senior Engineer* |

| HSO | Mr Rafarel, C.R. |

Culture Centre of Algae and Protozoa
36 Storey's Way
Cambridge
CB3 0DT
0223 (Cambridge) 61378

Senior Officer
Head of Subdivision

| SPSO | Mr George, E.A. |

Administration

EO	Miss Moxham, M.J.
C/tkr	Mr Yorke, D.A.
Clnr	Mrs Bleazard, D.I. (PT)
Clnr	Mrs Yorke, M.J. (PT)

DIVISION OF SCIENTIFIC SERVICES
Subdivision of Algal and Protozoan Culture

| SPSO | Mr George, E.A. *Head of Subdivision* |

PSO	Dr Belcher, J. Hilary
PSO	Dr Hibberd, D.J.
PSO	Dr Page, F.C.
PSO	Dr Swale, Erica, M.F.
SSO	Dr Morris, G.J.
SSO	Mr Clarke, K.J.
HSO	Mr Pennick, N.C.
SO	Mr Cann, J.P.
SO	Mrs Leeson, E.A.
ASO	Mrs Cann, S.F.
ASO	Mrs Coulson, G.E.
ASO	Miss Blower, A.E.
ASO	Mr Latham, N.D.

Publications

ABBOTT, A. (1978). Nutrient dynamics of ants. In: *Production ecology of ants and termites*, edited by M.V. Brian, 233–244. Cambridge: Cambridge University Press.

(ADAMS, R. &) HOOPER, M. D. (1978). *Nature day and night.* London: Kestrel.

ALLEN, S.E. (1978). *Chemistry in the Institute of Terrestrial Ecology.* Cambridge: Institute of Terrestrial Ecology.

BALL, D.F. (1978). Physiography, geology and soils of the grassland site at Llyn Llydaw. In: *Production ecology of British moors and montane grasslands*, edited by O.W. Heal & D.F. Perkins, 297–303. Berlin: Springer.

BALL, D.F. (1978). The soils of upland Britain. In: *The future of upland Britain*, edited by R.B. Tranter, 397–416. Reading: Centre for Agricultural Strategy.

(BAXTER, S.M. &) CANNELL, M.G.R. (1978). Branch development on leaders of *Picea sitchensis. Can. J. For. Res.*, **8**, 121–128.

BELCHER, J.H. (1978). A miniature battery-operated centrifuge. *Microscopy*, **33**, 278–279.

BENHAM, D.G. (& MELLANBY, K.) (1978). A device to exclude dust from rainwater samples. *Weather, Lond.*, **33**, 151–154.

BRASHER, S. & PERKINS, D.F. (1978). The grazing intensity and productivity of sheep in the grassland ecosystem. In: *Production ecology of British moors and montane grasslands*, edited by O.W. Heal & D.F. Perkins, 354–374. Berlin: Springer.

BRIAN M.V. *ed.* (1978). *Production ecology of ants and termites.* Cambridge: Cambridge University Press.

BRIAN, M.V. (& RIGBY, C.) (**1978**). The trophic eggs of *Myrmica rubra* L. *Insectes soc.*, **25**, 89–110.

(BROWN, D. &) ROTHERY, P. (1978). Randomness and local regularity of points in a plane. *Biometrika*, **65**, 115–122.

BUNCE, R.G.H. (& SMITH, R.S.). (1978). *An ecological survey of Cumbria.* Kendal: Cumbria County Council & Lake District Special Planning Board.

CALLAGHAN, T.V., COLLINS, N.J. (& CALLAGHAN, C.H.). (1978). Strategies of growth and population dynamics of tundra plants. 4. Photosynthesis, growth and reproduction of *Hylocomium splendens* and *Polytrichum commune* in Swedish Lapland. *Oikos*, **31**, 73–88.

CANNELL, M.G.R. (1978). Analysis of shoot apical growth of *Picea sitchensis* seedlings. *Ann. Bot.*, **42**, 1291–1303.

CANNELL, M.G.R. (& BOWLER, K.C.). (1978). Phyllotactic arrangements of needles on elongating conifer shoots: a computer simulation. *Can. J. For. Res.*, **8**, 138–141.

CANNELL, M.G.R. (& BOWLER, K.C.). (1978). Spatial arrangement of lateral buds at the time that they form on leaders of *Picea* and *Larix. Can. J. For. Res.*, **8**, 129–137.

CHAPMAN, S.B. & WEBB, N.R. (1978). The productivity of a Calluna heathland in southern England. In: *Production ecology of British moors and montane grasslands*, edited by O.W. Heal & D.F. Perkins, 247–262. Berlin: Springer.

(COCHRANE, L.A. &) FORD, E.D. (1978). Growth of a Sitka spruce plantation: analysis and stochastic description of the development of the branching structure. *J. appl. Ecol.*, **15**, 227–244.

(COLLINS, V.G., D'SYLVA, B.T. &) LATTER, P.M. (1978). Microbial populations in peat. In: *Production ecology of British moors and montane grasslands*, edited by O.W. Heal & D.F. Perkins, 94–112. Berlin: Springer.

COOKE, A.S. (1978). The action of sulphanilamide on shell deposition. *Br. Poultr. Sci.*, **19**, 267–272.

COOKE, A.S. (1978). Seasonal variation in sightings of foxes (*Vulpes vulpes*) at Woodwalton Fen. *Rep. Huntingdon. Fauna Flora Soc.*, 30th, 1977, 58–59.

COOKE, A.S. (1978). Shell structure of immature Roman snails (*Helix pomatia*) after exposure to *p,p'*-DDT. *Environ. Pollut.*, **17**, 31–37.

COOPER, J.I. (& TINSLEY, T.W.). (1978). Some epidemiological consequences of drastic ecosystem changes accompanying exploitation of tropical rain forest. *Terre Vie*, **32**, 221–240.

COOPER, J.I. (1978). *Virus and virus-like diseases of trees.* London: HMSO.

(CRUNDWELL, A.C. &) HILL, M.O. (1977). *Anoectangium warburgii*, a new species of moss from the British Isles. *J. Bryol.*, **9**, 435–440.

DALE, J. & HUGHES, R.E. (1978). Sheep population studies in relation to the Snowdonian environment. In: *Production ecology of British moors and montane grasslands*, edited by O. W. Heal & D.F. Perkins, 348–353. Berlin: Springer.

DAVIS, B.N.K. & JONES, P.E. (1978). The ground arthropods of some chalk and limestone quarries in England. *J. Biogeogr.*, **5**, 159–171.

DAVIS, B.N.K. (1978). Urbanisation and the diversity of insects. In: *Diversity of insect faunas*, edited by L.A. Mound & N. Waloff, 126–138. Oxford: Blackwell Scientific for Royal Entomological Society.

DEANS, J.D. & MILNE, R. (1978). An electrical recording soil moisture tensiometer. *Pl. Soil*, **50**, 509–513.

DUFFEY, E. (1978). Ecological strategies in spiders including some characteristics of species in pioneer and mature habitats. *Symp. zool. Soc. Lond.*, no. 42, 109–123.

(DUNCAN, J.S., REID, H.W.,) MOSS, R., (PHILLIPS, J.D. &) WATSON, A. (1978). Ticks, louping ill and red grouse on moors in Speyside, Scotland. *J. Wildl. Mgmt*, **42**, 500–505.

ELMES, G. W. (1978). A morphometric comparison of three closely related species of *Myrmica* (Formicidae), including a new species for England. *Syst. Entomol.*, **3**, 131–145.

(FLOOD, S.W. &) PERRING, F.H. (1978). *A handbook for local biological records centres.* Abbots Ripton: Biological Records Centre, Institute of Terrestrial Ecology.

FORD, E.D., (ROBARDS, A.W. & PINEY, M.D.). (1978). Influence of environmental factors on cell production and differentiation in the early wood of *Picea sitchensis. Ann. Bot.*, **42**, 683–692.

FOWLER, D. (1978). Dry deposition of SO_2 on agricultural crops. *Atmos. Environ.,* **12**, 369–373.

FRANKLAND, J.C., LINDLEY, D.K. (& SWIFT, M.J.). (1978). A comparison of two methods for the estimation of mycelial biomass in leaf litter. *Soil Biol. & Biochem.,* **10**, 323–333.

GEORGE, D.G. (& HEANEY, S.I.). (1978). Factors influencing the spatial distribution of phytoplankton in a small productive lake. *J. Ecol.,* **66**, 133–155.

GEORGE, E.A. (1978). The culture collection point of view. *Mitt. int. Verein. theor. angew. Limnol.,* **21**, 31–33.

GOOD, J.E.G., (CRAIGIE, I.,) LAST, F.T. & MUNRO, R.C. (1978). Conservation of amenity trees in the Lothian region of Scotland. *Biol. Conserv.,* **13**, 247–272.

(GORMAN, M.L.,) JENKINS, D. & HARPER, R.J. (1978). The anal scent tract of the otter (*Lutra lutra*). *J. Zool.,* **186**, 463–474.

GOSS-CUSTARD, J.D. (1978). Research on oystercatchers on the Exe estuary. *Devon Birds,* **31**, 45–50.

HARDING, P.T. & PLANT, R.A. (1978). A second record of *Cerambyx cerdo* L. (Coleoptera : Cerambycidae) from sub-fossil remains in Britain. *Entomologist's Gaz.,* **29**, 150–152.

HARRIS, M.P. (& HISLOP, J.R.G.). (1978). The food of young puffins *Fratercula arctica. J. Zool.,* **185**, 213–236.

HARRIS, M.P., (MORLEY, C. & GREEN, G.M.) (1978). Hybridization of herring and lesser black-backed gulls in Britain. *Bird Study,* **25**, 161–166.

HARRIS, M.P. (1978) Supplementary feeding of young puffins, *Fratercula arctica. J. Anim Ecol.,* **47**, 15–23.

HARRISON, A.F. (1978). Phosphorus cycles of forest and upland grassland ecosystems and some effects of land management practices. In: *Phosphorus in the environment,* edited by R. Porter & D.W. Fitzsimmons, 175–199. Amsterdam: Elsevier.

HEAL, O.W. (& SMITH, R.A.H.) (1978). Introduction and site description (Moor House). In: *Production ecology of British moors and montane grasslands,* edited by O.W. Heal & D.F. Perkins, 3–16. Berlin: Springer.

HEAL, O.W. & PERKINS, D.F., eds. (1978). *Production ecology of British moors and montane grasslands.* Berlin: Springer.

HEAL, O.W., LATTER, P.M. & HOWSON, G. (1978). A study of the rates of decomposition of organic matter .: *Production ecology of British moors and montane grasslands,* edited by O.W. Heal & D.F. Perkins, 136–159. Berlin: Springer.

HEATH, J. & SCOTT, D. (1977). Instructions for recorders. 2nd ed. Abbots Ripton: Biological Records Centre, Institute of Terrestrial Ecology.

HEATH, J. (1978). Lepidoptera–1977. *Rep. Huntingdon. Fauna Flora Soc.,* 30th, 1977, 26.

HELLIWELL, D.R. (1978). The assessment of landscape preferences. *Landscape Res.,* **3** (3) 15–17.

HELLIWELL, D.R. (1978). Floristic diversity in some central Swedish forests. *Forestry,* **51**, 149–161.

HELLIWELL, D.R. (1978). Forestry's long-term environmental role. In: *The future of upland Britain,* edited by R.B. Tranter, 108–113. Reading: Centre for Agricultural Strategy.

HELLIWELL, D.R. (1978). Survey and evaluation of wildlife on farmland in Britain: an 'indicator species' approach. *Biol. Conserv.,* **13**, 63–73.

HELLIWELL, D.R. & HARRISON, A.F. (1978). Variations in the growth of different seeds of *Acer pseudoplatanus* and *Betula verrucosa* grown on different soils. *Forestry,* **51**, 37–46.

HIBBERD, D.J. (1978). The fine structure of *Synura sphagnicola* (Korsh.) Korsh. (Chrysophyceae). *Br. phycol. J.,* **13**, 403–412.

HIBBERD, D.J. (1978). Possible phylogenetic value of the transitional helix in some chromophyte algal classes and some colourless protists. *BioSystems,* **10**, 115–116.

HILL, M.O. (1978). "A new flora of Gwynedd". *BSBI Welsh Bull.,* no. 28, 5.

HILL, M.O. (1978). Sphagnopsida. In: *The moss flora of Britain and Ireland,* by A.J.E. Smith, 30–78. Cambridge: Cambridge University Press.

HILL, M.O. (& JONES, E.W.). (1978). Vegetation changes resulting from afforestation of rough grazings in Caeo Forest, South Wales. *J. Ecol.,* **66**, 433–456.

HILL, M.O. & EVANS, D.F. (1978). The vegetation of upland Britain. In: *The future of upland Britain,* edited by R.B. Tranter, 436–447. Reading: Centre for Agricultural Strategy.

HOLDGATE, M.W. (& BEAMENT, J.W.L.) (1977). The ecological dilemma. In: *The encyclopaedia of ignorance,* edited by R. Duncan & M. Weston-Smith, 411–416. Oxford: Pergamon.

HOLDGATE, M.W. (1978). The application of ecological knowledge to land use planning. In: *The breakdown and restoration of ecosystems,* edited by M.W. Holdgate & M.J. Woodman, 451–464. London: Plenum.

HOLDGATE, M.W. & WOODMAN, M.J., eds. (1978). *The breakdown and restoration of ecosystems.* London: Plenum.

HOOPER, M.D. (1978). Changes in the landscape of woodland and hedgerow. In: *Conservation and agriculture,* edited by J.G. Hawkes, 89–94. London: Duckworth.

(JEFFERIES, D.J. &) ARNOLD, H.R. (1978). Mammal report for 1977. *Rep. Huntingdon. Fauna Flora Soc.,* 30th, 1977, 54–57.

JEFFERS, J.N.R. *Design of experiments.* (Statistical checklist 1). Cambridge: Institute of Terrestrial Ecology.

JEFFERS, J.N.R. (1978). The ecology of resource utilization. *J. oper. Res. Soc.,* **29**, 315–321.

JEFFERS, J.N.R. (1978). General principles for ecosystem definition and modelling. In: *The breakdown and restoration of ecosystems,* edited by M.W. Holdgate & M.J. Woodman, 85–101. London: Plenum.

JEFFERS, J.N.R. (1978). *Introduction to systems analysis: with ecological applications.* London: Arnold.

(JERMY, A.C.,) ARNOLD, H.R., FARRELL, L. & PERRING, F.H. (1978). *Atlas of ferns of the British Isles.* London: Botanical Society of the British Isles & British Pteridological Society.

JONES, H.E. & GORE, A.J.P. (1978). A simulation of production and decay in blanket bog. In: *Production ecology of British moors and montane grasslands,* edited by O.W. Heal & D.F. Perkins, 160–186. Berlin: Springer.

JONES, P.E. (1978). Phoresy and commensalism in British pseudoscorpions. *Proc. & Trans. Br. entomol. & natur. Hist. Soc.*, **11**, 90–96.

KING, M.L. (1978). Monks Wood butterflies. *Rep. Huntingdon. Fauna Flora Soc.*, 30th, 1977, 27.

KRUUK, H. (& HEWSON, R.). (1978). Spacing and foraging of otters (*Lutra lutra*) in a marine habitat. *J. Zool.*, **185**, 205–212.

KRUUK, H. (1978). Spatial organization and territorial behaviour of the European badger *Meles meles. J. Zool.*, **184**, 1–19.

LAST, F.T. (1978). Cryptogams and Scotland. *Trans. Proc. bot. Soc. Edinb.*, **42** (Suppl.) 99–124.

LAST, F.T. (1978). Effects of atmospheric pollutants on forests and natural plant assemblages. *Arboric. J.*, **3**, 324–340.

LAST, F.T. (1978). The right tree for the site. *Garden.* J.L.R. hort. Soc., **103**, 270–279.

LAST, F.T. (1978). Variety is the spice of life. *Ann. appl. Biol.*, **90**, 303–322.

LATTER, P.M. (1977). Decomposition of a moorland litter, in relation to *Marasmius androsaceous* and soil fauna. *Pedobiologia*, **17**, 418–427.

LATTER, P.M. & HOWSON, G. (1978). Studies on the microfauna of blanket bog with particular reference to Enchytraeidae. II. Growth and survival of *Cognettia sphagnetorum* on various substrates. *J. Anim. Ecol.*, **47**, 425–448.

LEAKEY, R.R.B., (CHANCELLOR, R.J. & VINCE-PRUE, D.). (1978). Regeneration from rhizome fragments of *Agropyron repens* (L.) Beauv. III. Effects of nitrogen and temperature on the development of dominance amongst shoots on multi-node fragments. *Ann. Bot.*, **42**, 197–204.

LEAKEY, R.R.B., (CHANCELLOR, R.J. & VINCE-PRUE, D.). (1978). Regeneration from rhizome fragments of *Agropyron repens* (L.) Beauv. IV. Effects of light on bud dormancy and the development of dominance amongst shoots on multi-node fragments. *Ann. Bot.*, **42**, 205–212.

LONGMAN, K.A. (1978). Control of flowering for forest tree improvement and seed production. *Scient. Hort.*, **30**, 1–10.

LONGMAN, K.A. (1978). Control of shoot extension and dormancy: external and internal factors. In: *Tropical trees as living systems*, edited by P.B. Tomlinson & M.H. Zimmermann, 465–495. Cambridge: Cambridge University Press.

LUTMAN, J. (1978). The role of slugs in an Agrostis-Festuca grassland. In: *Production ecology of British moors and montane grasslands*, edited by O.W. Heal & D.F. Perkins, 332–347. Berlin: Springer.

MAITLAND, P.S. (1978). *The biology of fresh waters.* London: Blackie.

MAITLAND, P.S. & MORRIS, K.H. (1978). A multi-purpose modular limnological sampler. *Hydrobiologia*, **59**, 187–195.

MARQUISS, M., NEWTON, I. (& RATCLIFFE, D.A.). (1978). The decline of the raven, *Corvus corax*, in relation to afforestation in southern Scotland and northern England. *J. appl. Ecol.*, **15**, 129–144.

MERRETT, P., ed. (1978). *Arachnology. 7th int. Congr. Arachnology, Exeter, 1977. Symp. zool. Soc. Lond.* no. 42. London: Academic Press for Zoological Society of London.

MILLER, G.R. & WATSON, A. (1978). Heather productivity and its relevance to the regulation of red grouse populations. In: *Production ecology of British moors and montane grasslands*, edited by O.W. Heal & D.F. Perkins, 277–285. Berlin: Springer.

MILLER, G.R. & WATSON, A. (1978). Territories and the food plant of individual red grouse. I. Territory size, number of mates and brood size compared with the abundance, production and diversity of heather. *J. Anim. Ecol.*, **47**, 293–305.

MILNER, C. (1978). Shetland ecology surveyed. *Geogrl Mag., Lond.*, **50**, 730–736.

MITCHELL, B. (& YOUNGSON, R.W.). *Teeth and age in Scottish red deer: a practical guide to age assessment.* Inverness: Red Deer Commission.

MITCHELL, B., PARISH, T. & CRISP, J.M. (1978). Weighing red deer in the field. *Deer*, **4**, 287–290.

MORIARTY, F. (1978). Starvation and growth in the gastropod *Planorbarius corneus* (L.). *Hydrobiologia*, **58**, 271–275.

MORIARTY, F. (1978). Terrestrial animals. In: *Principles of ecotoxicology*, edited by G. C. Butler, 169–186. London: Wiley.

MORRIS, G.J. & CLARKE, A. (1978). The cryopreservation of *Chlorella.* 4. Accumulation of lipid as a protective factor. *Arch. Microbiol.*, **119**, 153–156.

MORRIS, G.J. & CANNING, C.E. (1978). The cryopreservation of *Euglena gracilis. J. gen. Microbiol.*, **108**, 27–31.

MORRIS, M.G. (1978). *Polydrusus sericeus* (Schaller) in Wiltshire. *Entomologist's Rec. J. Var.*, **90**, 22.

MORRIS, M.G. (1978). *Polydrusus sericeus* (Schaller) (Col.: Curculionidae): an additional note. *Entomologist's Rec. J. Var.*, **90**, 55.

MOSS, R. (1977). The digestion of heather by red grouse during the spring. *Condor*, **79**, 471–477.

MURTON, R.K. (1978). The importance of photoperiod to artificial breeding in birds. *Symp. zool. Soc. Lond.*, no. 43, 7–29.

MURTON, R.K. (1978). Pesticides and wildlife: current ITE research. In: *Some aspects of research on pesticides*, 11–18. London: Natural Environment Research Council.

MURTON, R.K. (& KEAR, J.). (1978). Photoperiodism in waterfowl: phasing of breeding cycles and zoogeography. *J. Zool.*, **186**, 243–283.

NEWTON, I. (MEEK, E.R. & LITTLE, B.). (1978). Breeding ecology of the merlin in Northumberland. *Br. Birds.*, **71**, 376–398.

NEWTON, I., (1978). Feeding and development of sparrowhawk *Accipiter nisus* nestlings. *J. Zool.*, **184**, 465–488.

NEWTON, I. (& BOGAN, J.). (1978). The role of different organo-chlorine compounds in the breeding of British sparrowhawks. *J. appl. Ecol.*, **15**, 105–116.

OSBORN, D. (1978). The alpha adrenergic receptor mediated increase in guinea-pig liver glycogenolysis. *Biochem. Pharmac.*, **27**, 1315–1320.

OSBORN, D. (1978). A naturally occurring cadmium and zinc binding protein from the liver and kidney of *Fulmarus glacialis*, a pelagic north Atlantic seabird. *Biochem. Pharmac.*, **27**, 822–824.

PAGE, F.C. (1978). *Acrasis rosea* and the possible relationship between Acrasida and Schizopyrenida. *Arch. Protistenk.*, **120**, 169–181.

PAGE, F.C. (1978). An electron-microscopical study of *Thecamoeba proteoides* (Gymnamoebia) intermediate between Thecamoebidae and Amoebidae. *Protistologica*, **14**, 77–86.

PELHAM, J. & MASON, P. (1978). Aseptic cultivation of sapling trees for studies of nutrient responses with particular reference to phosphate. *Ann. appl. Biol.*, **88**, 415–419.

PENNICK, N.C., CLARKE, K.J. & BELCHER, J.H. (1978). Studies of the external morphology of *Pyramimonas*. 1. *P. orientalis* and its allies in culture. *Arch. Protistenk.*, **120**, 304–311.

PENNICK, N.C. (1978). Studies of the external morphology of *Pyramimonas*. 5. *P. amylifera* Conrad, with 13 figures. *Arch. Protistenk.*, **120**, 142–147.

PERKINS, D.F. (1978). The distribution and transfer of energy and nutrients in the Agrostis-Festuca grassland ecosystem. In: *Production ecology of British moors and montane grasslands*, edited by O.W. Heal & D.F. Perkins, 375–395. Berlin: Springer.

PERKINS, D.F., JONES, V., MILLAR, R.O. & NEEP, P. (1978). Primary production, mineral nutrients and litter decomposition in the grassland ecosystem. In: *Production ecology of British moors and montane grasslands*, edited by O.W. Heal and D.F. Perkins 304–331. Berlin: Springer.

PERKINS, D.F. (1978). Snowdonia grassland: introduction, vegetation and climate. In: *Production ecology of British moors and montane grasslands*, edited by O.W. Heal & D.F. Perkins, 289–296. Berlin: Springer.

PERRING, F.H. (1977). The role of natural history collections in preparing distribution maps. *Museums J.*, **77**, 133.

PICOZZI, N. (1978). Dispersion, breeding and prey of the hen harrier *Circus cyaneus* in Glen Dye, Kincardineshire. *Ibis*, **120**, 498–509.

PLANT, R.A. & HARDING, P.T. (1978). Sub-fossil remains of *Cerambyx cerdo* L. (Col., Cerambycidae) at Ramsey Heights, Huntingdonshire. *Rep. Huntingdon. Fauna Flora Soc.*, 30th, 1977, 40–41.

(RAWES, M. &) HEAL, O.W. (1978). The blanket bog as part of a Pennine moorland. In: *Production ecology of British moors and montane grasslands*, edited by O.W. Heal & D.F. Perkins, 224–243. Berlin: Springer.

READING, C.J. & McGRORTY, S. (1978). Seasonal variations in the burying depth of *Macoma balthica* (L.) and its accessibility to wading birds. *Estuar. & coast. mar. Sci.*, **6**, 135–144.

(REID, H.W., DUNCAN, J.S., PHILLIPS, J.D.P.,) MOSS, R. & WATSON, A. (1978). Studies on louping-ill virus (Flavivirus group) in wild red grouse (*Lagopus lagopus scoticus*). *J. Hyg., Camb.*, **81**, 321.

SATCHELL, J.E. (& STONE, D.A.). (1977). Colonisation of pulverized fuel ash sites by earthworms. *Publ. Cent. pirenaico Biol. exp.*, **9**, 59–74.

SATCHELL, J.E. (1977). Earthworms—the trombones of the grave. In: *Soil organisms as components of ecosystems*, edited by U. Lohm & T. Persson, 598–603. Stockholm: Swedish Natural Science Research Council.

SATCHELL, J.E. (1978). Ecology and environment in the United Arab Emirates. *J. arid Environments*, **1**, 201–226.

SAVORY, C.J. (1978). Food consumption of the red grouse in relation to the age and productivity of heather. *J. Anim. Ecol.*, **47**, 269–282.

SHEAIL, J. (1978). Rabbits and agriculture in post-medieval England. *J. hist. Geog.*, **4**, 343–355.

STAINES, B.W. (1978). The dynamics and performance of a declining population of red deer (*Cervus elaphus*). *J. Zool.*, **184**, 403–419.

STAINES, B.W. (1978). Factors affecting the seasonal distribution of red deer (*Cervus elaphus*) at Glen Dye, north-east Scotland. *Ann. appl. Biol.*, **88**, 347.

STAINES, B.W. & CRISP, J.M. (1978). Observations on food quality in Scottish red deer (*Cervus elaphus*) as determined by chemical analysis of the rumen contents. *J. Zool.*, **185**, 253–259.

THOMAS, D.I. & GOOD, J.E.G. (1978). Future possibilities for fuel cropping in upland Britain. In: *The future of upland Britain*, edited by R.B. Tranter, 366–373. Reading: Centre for Agricultural Strategy.

WARD, P. (1978). The role of the crop among red-billed queleas *Quelea quelea*. *Ibis*, **120**, 333–337.

WATSON, A. (1978). Grouse research news. *Shooting times*, (4993) 15.

WATSON, A. (1978). Red grouse. *Ann. J. Br. Field Sports Soc.*, **1**, 26–27.

WATSON, A. & STAINES, B. W. (1978). Differences in the quality of wintering areas used by male and female red deer (*Cervus elaphus*) in Aberdeenshire. *J. Zool.*, **186**, 544–550.

WEBB, N.R. (1977). Observations on *Steganacarus magnus*: general biology and life cycle. *Acarologia*, **19**, 686–696.

WELCH, J.M. & POLLARD, E. (1978). Review of current literature on the ecology and exploitation of the edible snail *Helix pomatia*. *Malacol. Rev.*, **10**, 1–6.

WELCH, R.C. (1978). Changes in the distribution of the nests of *Formica rufa* L. (Hymenoptera, Formicidae) at Blean Woods National Nature Reserve, Kent, during the decade following coppicing. *Insectes soc.*, **25**, 173–186.

WELCH, R.C. (1978). Insects recorded by Professor C.C. Babington in Monks Wood and other Huntingdonshire localities, 1828–1836. *Rep. Huntingdon. Fauna Flora Soc.*, 30th, 1977, 42–43.

WELLS, T.C.E. (1978). Botanical notes. *Rep. Huntingdon. Fauna Flora Soc.*, 30th, 1977, 3–6.

WELLS, T.C.E. (1978). Establishment of species-rich chalk grassland on previously arable land. *J. Sports Turf Res. Inst.*, no. 53, 1977, 105.

WELLS, T.C.E. & SHEAIL, J. (1978). The Marchioness of Huntley's botanical records for Huntingdonshire (v.c. 31). *Rep. Huntingdon. Fauna Flora Soc.*, 30th, 1977, 7–24.

WELLS, T.C.E. & SHEAIL, J. (1978). Plant records from Cambridgeshire (v.c. 29) in the Marchioness of Huntley's herbarium. *Nature Cambs.*, **21**, 38–39.

Merlewood Research and Development Papers
72.   HOWARD, P.J.A. (1977). The survey and classification of upland areas for land management studies—a review.

73. THOMAS, D.I. (1977). The energy outlook.

74. HOWARD, P.J.A. (1977). Numerical classification and cluster analysis in ecology: a review.

75. BISHOP, I. (1978). Land use in rural Cumbria—a linear programming model.

Reports and other publications
INSTITUTE OF TERRESTRIAL ECOLOGY (1978). *Annual report 1977*. Cambridge: Institute of Terrestrial Ecology.

INSTITUTE OF TERRESTRIAL ECOLOGY (1978). *Overlays of environmental and other factors for use with Biological Records Centre distribution maps*. Cambridge: Institute of Terrestrial Ecology.